T0135955

Overcoming the failure of the classical generalized interior-point regularity conditions in convex optimization.

Applications of the duality theory to enlargements of maximal monotone operators

Ernö Robert Csetnek

Logos Verlag Berlin

λογος

Bibliografische Information der Deutschen Nationalbibliothek

Die Deutsche Nationalbibliothek verzeichnet diese Publikation in der
Deutschen Nationalbibliografie; detaillierte bibliografische Daten sind
im Internet über http://dnb.d-nb.de abrufbar.

ISBN 978-3-8325-2503-3

Logos Verlag Berlin GmbH
Comeniushof, Gubener Str. 47,
10243 Berlin
Tel.:   +49 (0)30 / 42 85 10 90
Fax: +49 (0)30 / 42 85 10 92
http://www.logos-verlag.de

Bibliographical description

Ernö Robert Csetnek

Overcoming the failure of the classical generalized interior-point regularity conditions in convex optimization. Applications of the duality theory to enlargements of maximal monotone operators

Dissertation, **108** pages, Chemnitz University of Technology, Faculty of Mathematics, 2009

Abstract

The aim of this work is to present several new results concerning duality in scalar convex optimization, the formulation of sequential optimality conditions and some applications of the duality to the theory of maximal monotone operators.

After recalling some properties of the classical generalized interiority notions which exist in the literature, we give some properties of the quasi interior and quasi-relative interior, respectively. By means of these notions we introduce several generalized interior-point regularity conditions which guarantee Fenchel duality. By using an approach due to Magnanti, we derive corresponding regularity conditions expressed via the quasi interior and quasi-relative interior which ensure Lagrange duality. These conditions have the advantage to be applicable in situations when other classical regularity conditions fail. Moreover, we notice that several duality results given in the literature on this topic have either superfluous or contradictory assumptions, the investigations we make offering in this sense an alternative.

Necessary and sufficient sequential optimality conditions for a general convex optimization problem are established via perturbation theory. These results are applicable even in the absence of regularity conditions. In particular, we show that several results from the literature dealing with sequential optimality conditions are rediscovered and even improved.

The second part of the thesis is devoted to applications of the duality theory to enlargements of maximal monotone operators in Banach spaces. After establishing a necessary and sufficient condition for a bivariate infimal convolution formula, by employing it we equivalently characterize the ε-enlargement of the sum of two maximal monotone operators. We generalize in this way a classical result concerning the formula for the ε-subdifferential of the sum of two proper, convex and lower semicontinuous functions. A characterization of fully enlargeable monotone operators is also provided, offering an answer to an open problem stated in the literature. Further, we give a regularity condition for the weak*-closedness of the sum of the images of enlargements of two maximal monotone operators.

The last part of this work deals with enlargements of positive sets in SSD spaces. It is shown that many results from the literature concerning enlargements of maximal monotone operators can be generalized to the setting of Banach SSD spaces.

Keywords

conjugate functions; quasi interior; quasi-relative interior; conjugate duality; Fenchel and Lagrange duality; convex optimization; separation theorems; regularity conditions; perturbation functions; ε-subdifferential; (convex) subdifferential; sequential optimality conditions; maximal monotone operators; Fitzpatrick function; representative function; enlargement; positive sets; SSD spaces

Acknowledgements

I wish to express my gratitude to my advisor, Prof. Dr. Gert Wanka, for proposing me this research topic and for his constant support and assistance during my doctoral study.

Special thanks go to Dr. Radu Ioan Boţ for the continuous supervision of my research work, stimulating discussions, advice and friendliness. It has been a privilege to study under his guidance.

I would like to thank Dr. Sorin-Mihai Grad for his help and support during these years.

I also want to thank the Free State of Saxony for granting me with a graduate fellowship in the period 04/2006-03/2009.

I am grateful to the Faculty of Mathematics, Chemnitz University of Technology, for providing me a good research environment.

Many thanks to my family for love, understanding and encouragements.

Last, but not least, my sincere thanks to my wife Minodora for love, patience, and for believing in me all the time.

Contents

Chapter 1

Introduction

The *simplex method* published by DANTZIG in 1947 and the duality theorem (explicitly given for the first time in 1951 by GALE, KUHN AND TUCKER [66]) have proved to be important steps in *linear optimization*, due to their robustness and efficiency for solving various problems appearing in operations research, business, economics and engineering.

Soon it was realized that in practice often one has to deal with optimization problems with the function which has to be minimized (or maximized) being convex, and not necessarily linear. This fact along with the increasing interest of mathematicians in the *calculus of variations* motivated an intensive study of convex sets and convex functions. We mention here the pioneering works of FENCHEL [64], BRØNDSTED [40], MOREAU [104, 105] and ROCKAFELLAR [120, 121], which are the cornerstones of the *convex analysis*, including investigations on the theory of convex functions, conjugate functions and duality in convex optimization. For a comprehensive study of *convex analysis* in finite-dimensional spaces we refer to the monographs of BORWEIN AND LEWIS [15], HIRIART-URRUTY AND LEMARÉCHAL [75–77] and ROCKAFELLAR [122], while for the infinite-dimensional case we mention the works due to BORWEIN AND VANDERWERFF [17], BOŢ [21], BOŢ, GRAD AND WANKA [36], EKELAND AND TEMAM [62] and ZĂLINESCU [149] (see also [123]).

To a primal convex optimization problem one can associate, by means of conjugate functions, a dual optimization problem, for which *weak duality* holds, that is the optimal objective value of the dual is less than or equal to the optimal objective value of the primal problem. Let us mention that the two duality approaches with the greatest resonance in the literature are the so-called *Fenchel* and *Lagrange duality*, respectively. Actually, the duality approach based on conjugate functions can be studied from a more general point of view, by means of the *perturbation theory* (we refer to [62, 149] for more on this approach). An important challenge in duality theory is to find conditions which ensure *strong duality*, namely the case when the optimal objective values of the two problems are equal and the dual has an optimal solution. This issue was solved by introducing several so-called *regularity conditions* guaranteeing strong duality.

Let us mention that several results from the theory of conjugate duality have been successfully applied in mathematical economics, optimal control, mechanics, numerical analysis, variational analysis, support vector machines and the list can be continued.

The present work has been developed towards two main directions. In the first part we introduce by means of generalized interiority notions some new regularity conditions guaranteeing Fenchel and Lagrange duality. These conditions are useful to overcome the situation when the classical regularity conditions given in the literature fail. Moreover, establishing regularity conditions guaranteeing strong duality is important in order to be able to derive necessary and sufficient *optimality conditions*. On the other hand, we show that a sequential form of optimality conditions for different classes of optimization problems can be provided even in the absence of any regularity condition.

The second part of the thesis is dedicated to applications of the duality theory to enlargements of monotone operators. Since 2001, when the *Fitzpatrick function* associated to a maximal monotone operator was rediscovered, conjugate duality plays a significant role in the theory of maximal monotone operators, offering in many situations the possibility to reduce questions on monotone operators to questions on convex functions. We underline this connection by several applications of the duality theory to enlargements of maximal monotone operators in Banach spaces.

1.1   A description of the contents

In the following we give a description of the contents of this thesis, underlining its most important results. In the last part of the introduction we include a section with preliminary notions and results which makes this manuscript self-contained.

The second chapter is devoted to the study of strong duality results in infinite-dimensional scalar convex optimization. In the first section we revisit some properties of several generalized interiority notions from the literature, like the *algebraic interior*, *relative algebraic interior* and *strong quasi-relative interior*. Then we focus our attention on the notions of *quasi interior* and *quasi-relative interior*, the later being introduced by BORWEIN AND LEWIS (cf. [14]). The main tool which is often used in deriving strong duality results in convex optimization is the existence of separation theorems. This in the reason why a special attention is paid to establishing useful separation theorems by means of the quasi interior and quasi-relative interior of convex sets. The next section deals with Fenchel duality. After recalling the classical generalized interior-point regularity conditions given in the literature in order to overcome the duality gap between a primal (Fenchel-type) convex optimization problem and its Fenchel dual, we introduce some new ones expressed with the help of the notions of quasi interior and quasi-relative interior, respectively. These conditions turn out to be sufficient for strong duality. A very interesting approach due to MAGNANTI (cf. [93]) offers a link between Fenchel duality and Lagrange duality. This is presented in the last section of the second chapter and we derive in this way corresponding strong duality results between the primal optimization problem with geometric and cone constraints and its Lagrange dual problem. The strong duality results introduced by means of the quasi interior and quasi-relative interior offer an alternative for the situation when the classical strong duality results from the literature cannot be applied. Several examples illustrate the usefulness of these new duality statements. We conclude the chapter with a comment on different strong duality theorems given in the literature that employ the quasi-relative interior, which turn out to have either superfluous, or contradictory assumptions. The investigations we make are useful to overcome this drawback.

In the first section of the third chapter we derive necessary and sufficient sequential optimality conditions for the general optimization problem

$$(P_\Phi) \quad \inf_{x \in X} \Phi(x, 0),$$

where $\Phi : X \times Y \to \overline{\mathbb{R}}$ is a proper, convex and lower semicontinuous function such that $0 \in \mathrm{pr}_Y(\mathrm{dom}\,\Phi)$ and X, Y are Banach spaces with X being reflexive. These sequential characterizations of optimal solutions have the advantage to be applicable even in the case when no regularity condition is fulfilled. In the last three sections of this chapter we particularize the general results, rediscovering, and in many situations, even improving some sequential optimality conditions given in the literature.

For a particular choice of the function Φ we derive in Section 3.2 sequential optimality conditions for the optimization problem with the objective function being the sum of a proper, convex and lower semicontinuous function with the composition of another proper, convex and lower semicontinuous function with a continuous linear operator. The sequential generalizations of the Pshenichnyi-Rockafellar Lemma given by JEYAKUMAR AND WU in [86, Theorem 3.3 and Corollary 3.5]) follow as particular cases of the general theory. For an appropriate choice of the function Φ we give in Section 3.3 qualification free necessary and sufficient sequential optimality conditions for the optimization problem with geometric and cone constraints, improving a result given by THIBAULT in [138, Theorem 4.1]). In the last section of this chapter we provide different sequential optimality conditions for composed convex optimization problems. We show that also in this case some sequential characterizations of subgradients given by THIBAULT in [138] follow as particular cases of the general theory developed in Section 3.1.

In the fourth chapter of the thesis we present some applications of conjugate duality to enlargements of maximal monotone operators in Banach spaces. In Section 4.1 we introduce a closedness-type regularity condition which turns out to be necessary and sufficient in order to have a so-called *bivariate infimal convolution formula*. In the next section we revisit the most important notions and results concerning monotone operators and their enlargements, including the properties of the so-called *Fitzpatrick function*, which establishes the connection between different elements of convex analysis and the theory of monotone operators. By using the conditions given for the bivariate infimal convolution formula, we equivalently characterize in Section 4.3 the ε-enlargement of the sum of two maximal monotone operators. We extend by this approach a classical result regarding an equivalent characterization of the ε-subdifferential of the sum of two proper, convex and lower semicontinuous functions. BURACHIK AND IUSEM posed in [44] an open problem concerning the characterization of the maximal monotone operators $S : X \rightrightarrows X^*$ (X is a Banach space) which are fully enlargeable by S^{se}, the smallest element belonging to a special family of enlargements associated to S. An answer to this problem is provided in Section 4.4. In the last section of this chapter we introduce a weak regularity condition which guarantees the weak*-closedness of the set $S_{h_S}(\varepsilon_1, x) + T_{h_T}(\varepsilon_2, x)$, where $\varepsilon_1, \varepsilon_2 \geq 0$, $S, T : X \rightrightarrows X^*$ are two maximal monotone operators with representative functions h_S and h_T, respectively, while X and Y are Banach spaces. This is achieved by giving a preliminary result that ensures the weak*-closedness of the sum of two convex and weak*-closed sets, which are actually sublevel sets of some functions with certain properties. In case X is a reflexive Banach space, or X is Banach and S, T are of Gossez type (D), we improve in this way a result given by GARCÍA, LASSONDE AND REVALSKI in [67, Theorem 3.7 (1)].

In the last chapter we study the theory of enlargements of monotone operators from a more abstract, though systematic way. STEPHEN SIMONS introduced in [129] the notion of *positive set* with respect to a quadratic form q defined on a so-called *symmetrically self-dual Banach space (Banach SSD space)*, as an extension of the notion of a monotone set in Banach spaces. Let us notice that the term *Simons space* is already used in the community when referring to the notion of *Banach SSD space* (see [23, 110]). A number of known results coming from the theory of monotone operators has been successfully generalized to this framework. In analogy to the enlargement of a monotone operator we introduce and study the notion of enlargement of a positive set in SSD spaces. In Section 5.1 we investigate the algebraic properties of this notion, like convexity, transportation formula, etc. We also associate to a positive set A a family of enlargements $\mathbb{E}(A)$ for which we provide, in case A is a maximally q-positive set, the smallest and the biggest element with respect to the partial ordering inclusion relation of the graphs. In Section 5.2 we give some topological properties of enlargements of positive sets in the framework of Banach

SSD spaces. For $\mathbb{E}_c(A)$, the subfamily of $\mathbb{E}(A)$ containing the enlargements of A having a closed graph, we point out, in case A is maximally q-positive, the smallest and the biggest element with respect to the partial ordering inclusion relation of the graphs. A one-to-one correspondence is established between this subfamily and $\mathcal{H}(A)$, the set of so-called *representative functions of A*. We also show that the smallest and the biggest elements of $\mathcal{H}(A)$ are nothing else than two functions considered by SIMONS in [131]. We close the chapter by giving a characterization of the additive enlargements in $\mathbb{E}_c(A)$, in case A is a maximally q-positive set, which turns out to be helpful when showing the existence of enlargements having this property. In this way we extend to (Banach) SSD spaces several results given by BURACHIK AND SVAITER in [50, 51, 134] for enlargements of maximal monotone operators.

1.2   Preliminary notions and results

For the functional analysis tools considered in this work we refer to the monographs of FABIAN et al. [63], HOLMES [78] and RUDIN [127]. We emphasize that all the vector spaces mentioned in this work are considered over the real field.

Throughout this thesis we denote by $\mathbb{N}$ the set of positive integers $\{1, 2...\}$, $\mathbb{Z}$ the set of integer numbers, $\mathbb{R}$ the set of real numbers and $\mathbb{R}_+$ the set of non-negative real numbers. Let us mention that we denote by $\mathcal{R}$ the usual topology on $\mathbb{R}$. We consider also $\overline{\mathbb{R}} = \mathbb{R} \cup \{\pm\infty\}$ the extended real line. By $\mathbb{R}^n$ $(n \in \mathbb{N})$ we denote the n-dimensional space and by $\mathbb{R}^n_+$ the non-negative orthant of $\mathbb{R}^n$, that is $\mathbb{R}^n_+ = \{(x_1, ..., x_n) \in \mathbb{R}^n : x_i \geq 0 \ \forall i = \overline{1, n}\}$. We identify $\mathbb{R}^1$ with $\mathbb{R}$ and similarly $\mathbb{R}^1_+$ with $\mathbb{R}_+$.

Consider X a separated locally convex space and X^* its topological dual space. We denote by $w(X, X^*)$ $(w(X^*, X))$ the weak topology on X induced by X^* (the weak* topology on X^* induced by X). When there is no danger of confusion, the notation w (w^*) is used. For a non-empty set $U \subseteq X$, we denote by $\mathrm{co}(U), \mathrm{cone}(U), \mathrm{coneco}(U), \mathrm{aff}(U), \mathrm{lin}(U), \mathrm{int}(U), \mathrm{ri}(U), \mathrm{cl}(U)$, its *convex hull, conic hull, convex conic hull, affine hull, linear hull, interior, relative interior*, and *closure*, respectively. We have $\mathrm{cone}(U) = \cup_{t \geq 0} tU$ and if $0 \in U$ then obviously $\mathrm{cone}(U) = \cup_{t > 0} tU$. The set $\mathrm{ri}(U)$ is the interior of U relative to $\mathrm{cl}\left(\mathrm{aff}(U)\right)$. In finite-dimensional spaces, $\mathrm{ri}(U)$ is the classical relative interior, that is the interior of U relative to $\mathrm{aff}(U)$. Let us consider $V \subseteq X$ another non-empty set. By $U + V$ we denote the usual *Minkowski sum* of the sets $U, V \subseteq X$, that is $U + V = \{u + v : u \in U, v \in V\}$, while for $\alpha \in \mathbb{R}$, $\alpha U = \{\alpha x : x \in U\}$. By convention we take $U + \emptyset = \emptyset + U = \emptyset + \emptyset = \alpha\emptyset = \emptyset$. The following property will be used several times in Chapter 2: if U is convex then

$$\mathrm{coneco}(U \cup \{0\}) = \mathrm{cone}(U). \tag{1.1}$$

We denote by $\langle x^*, x \rangle$ the value of the continuous linear functional $x^* \in X^*$ at $x \in X$. Consider the *identity function* on X, $\mathrm{id}_X : X \to X, \mathrm{id}_X(x) = x$ for all $x \in X$. For a function $f : U \times V \to \overline{\mathbb{R}}$ we denote by $f^\top$ the *transpose* of f, namely the function $f^\top : V \times U \to \overline{\mathbb{R}}, f^\top(v, u) = f(u, v)$ for all $(v, u) \in V \times U$. Let us mention also the *projection operator* $\mathrm{pr}_U : U \times V \to U, \mathrm{pr}_U(u, v) = u$ for all $(u, v) \in U \times V$. The *indicator function* of U, denoted by δ_U, is defined as $\delta_U : X \to \overline{\mathbb{R}}$,

$$\delta_U(x) = \begin{cases} 0, & \text{if } x \in U, \\ +\infty, & \text{otherwise.} \end{cases}$$

The *support function* of U, $\sigma_U : X^* \to \overline{\mathbb{R}}$ is defined by $\sigma_U(x^*) = \sup_{x \in U} \langle x^*, x \rangle$ for all $x^* \in X^*$. We say that the function $f : X \to \overline{\mathbb{R}}$ is convex if

$$\forall x, y \in X, \ \forall t \in [0, 1] \ : f(tx + (1 - t)y) \leq tf(x) + (1 - t)f(y), \tag{1.2}$$

with the conventions $(+\infty) + (-\infty) = +\infty$, $0 \cdot (+\infty) = +\infty$ and $0 \cdot (-\infty) = 0$ (see [149]). We consider $\operatorname{dom} f = \{x \in X : f(x) < +\infty\}$ the *domain* of f and $\operatorname{epi} f = \{(x,r) \in X \times \mathbb{R} : f(x) \leq r\}$ its *epigraph*. Moreover, we denote by $\widehat{\operatorname{epi}}(f) = \{(x,r) \in X \times \mathbb{R} : (x,-r) \in \operatorname{epi} f\}$, the *symmetric* of $\operatorname{epi} f$ with respect to the x-axis. For a given real number α, $f - \alpha : X \to \overline{\mathbb{R}}$ is the function defined by $(f - \alpha)(x) = f(x) - \alpha$ for all $x \in X$. We call f *proper* if $\operatorname{dom} f \neq \emptyset$ and $f(x) > -\infty$ for all $x \in X$. By $\operatorname{cl} f$ we denote the *lower semicontinuous hull* of f, namely the function whose epigraph is the closure of $\operatorname{epi} f$ in $X \times \mathbb{R}$, that is $\operatorname{epi}(\operatorname{cl} f) = \operatorname{cl}(\operatorname{epi} f)$. We consider also $\operatorname{co} f$, the *convex hull* of f, which is the greatest convex function majorized by f. For $x \in X$ such that $f(x) \in \mathbb{R}$ we define the ε-*sudifferential* of f at x, where $\varepsilon \geq 0$, by

$$\partial_\varepsilon f(x) = \{x^* \in X^* : f(y) - f(x) \geq \langle x^*, y - x \rangle - \varepsilon \ \forall y \in X\}.$$

If $f(x) \in \{\pm\infty\}$ we take by convention $\partial_\varepsilon f(x) = \emptyset$. The set $\partial f(x) = \partial_0 f(x)$ is the classical (convex) *subdifferential* of f at x. The ε-subdifferential, introduced in [41], plays an important role in convex analysis, having significant theoretical and practical applications. The ε-subdifferential of f is an "enlargement" of its subdifferential, in the sense that $\partial f(x) \subseteq \partial_\varepsilon f(x)$ for all $x \in X$ and $\varepsilon \geq 0$. Let us mention that if f is proper, convex, continuous at $x_0 \in \operatorname{dom} f$ and Gâteaux differentiable at x_0, then $\partial f(x_0) = \{\nabla f(x_0)\}$ (cf. [149, Corollary 2.4.10 and Theorem 2.4.4(i)]). The following fact underlines the usefulness of the subdifferential: if f is proper then for $a \in \operatorname{dom} f$ we have the relation

$$\inf_{x \in X} f(x) = f(a) \Leftrightarrow 0 \in \partial f(a).$$

The ε-*normal set* of U at $x \in X$ is defined by $N_U^\varepsilon(x) = \partial_\varepsilon \delta_U(x)$, that is $N_U^\varepsilon(x) = \{x^* \in X^* : \langle x^*, y - x \rangle \leq \varepsilon \ \forall y \in U\}$ when $x \in U$, and $N_U^\varepsilon(x) = \emptyset$ if $x \notin U$. The *normal cone* of U at $x \in X$ is $N_U(x) = N_U^0(x)$, that is $N_U(x) = \{x^* \in X^* : \langle x^*, y - x \rangle \leq 0 \ \forall y \in U\}$, if $x \in U$ and $N_U(x) = \emptyset$ otherwise.

The *Fenchel-Moreau conjugate* of f is the function $f^* : X^* \to \overline{\mathbb{R}}$ defined by

$$f^*(x^*) = \sup_{x \in X} \{\langle x^*, x \rangle - f(x)\} \ \forall x^* \in X^*.$$

It holds $\delta_U^* = \sigma_U$. We mention here some important properties of conjugate functions. We have the so-called *Young-Fenchel inequality*

$$f^*(x^*) + f(x) \geq \langle x^*, x \rangle \ \forall x \in X \ \forall x^* \in X^*.$$

The *Fenchel-Moreau Theorem* is used several times throughout this work. This states that if f is proper, then f is convex and lower semicontinuous if and only if $f^{**} = f$ (see [62, 149]). Moreover, if f is convex and $(\operatorname{cl} f)(x) > -\infty$ for all $x \in X$, then $f^{**} = \operatorname{cl} f$ (cf. [149, Theorem 2.3.4]).

The following characterizations of the subdifferential and ε-sudifferential of a proper function f at $x \in \operatorname{dom} f$ by means of conjugate functions will be useful (see [62, 149]):

$$x^* \in \partial f(x) \Leftrightarrow f(x) + f^*(x^*) = \langle x^*, x \rangle \Leftrightarrow f(x) + f^*(x^*) \leq \langle x^*, x \rangle$$

and, respectively,

$$x^* \in \partial_\varepsilon f(x) \Leftrightarrow f(x) + f^*(x^*) \leq \langle x^*, x \rangle + \varepsilon.$$

In case $f : X \to \overline{\mathbb{R}}$ is a proper function and $a \in \operatorname{dom} f$, the epigraph of f^* can be represented as follows

$$\operatorname{epi} f^* = \bigcup_{\varepsilon \geq 0} \{(x^*, \langle x^*, a \rangle + \varepsilon - f(a)) : x^* \in \partial_\varepsilon f(a)\}. \qquad (1.3)$$

This formula, which is an easy consequence of the definitions above, describes the epigraph of a conjugate function in terms of the ε-subdifferential of the function and will play an important role in the proof of the main results of Chapter 3. It was stated in [81], where the function f was considered convex and lower semicontinuous, however the formula is valid even without these hypotheses.

The following version of the *Brøndsted-Rockafellar Theorem* (see [41]) was proved in [138] and will be used for providing sequential optimality conditions written in terms of the subdifferentials of the functions involved (see Section 3.1).

Theorem 1.1 *(Brøndsted-Rockafellar Theorem [41,138]) Let $(X, \|\cdot\|)$ be a Banach space, $f : X \to \overline{\mathbb{R}}$ be a proper, convex and lower semicontinuous function and $a \in \operatorname{dom} f$. Then for every $\varepsilon > 0$ and for every $x^* \in \partial_\varepsilon f(a)$ there exist $x_\varepsilon \in X$ and $x_\varepsilon^* \in \partial f(x_\varepsilon)$ such that*

$$\|x_\varepsilon - a\| \leq \sqrt{\varepsilon}, \ \|x_\varepsilon^* - x^*\|_* \leq \sqrt{\varepsilon} \ and \ |f(x_\varepsilon) - \langle x_\varepsilon^*, x_\varepsilon - a\rangle - f(a)| \leq 2\varepsilon,$$

where $\|\cdot\|_*$ is the dual norm on X^*.

Having $f, g : X \to \overline{\mathbb{R}}$ two functions we consider their *infimal convolution*, namely the function denoted by $f\square g : X \to \overline{\mathbb{R}}$, $f\square g(x) = \inf_{u \in X}\{f(u) + g(x - u)\}$ for all $x \in X$. We say that the infimal convolution is *exact* at $x \in X$ if the infimum in its definition is attained. Moreover, $f\square g$ is said to be *exact* if it is exact at every $x \in X$. We refer to [105,133] for more properties of the infimal convolution operation.

Let us also note that everywhere within this work we write min (max) instead of inf (sup) when the infimum (supremum) is attained.

Consider Y another separated locally convex space. For a function $h : X \to Y$ we denote by $h(U) = \{h(u) : u \in U\}$ the *image* of the set $U \subseteq X$ through h, while for $D \subseteq Y$ we use the notation $h^{-1}(D) = \{x \in X : h(x) \in D\}$. Given a continuous linear mapping $A : X \to Y$, its *adjoint operator*, $A^* : Y^* \to X^*$ is defined by $\langle A^*y^*, x\rangle = \langle y^*, Ax\rangle$ for all $y^* \in Y^*$ and $x \in X$. Consider also a non-empty *convex cone* $C \subseteq Y$ (that is cone$(C) \subseteq C$ and $C + C \subseteq C$) and $C^* = \{y^* \in Y^* : \langle y^*, y\rangle \geq 0 \ \forall y \in C\}$ its *positive dual cone*. By C^- we denote the *negative dual cone* of C, that is $C^- = -C^*$. Let $\leq_C$ be the partial ordering induced by C on Y, defined as $y_1 \leq_C y_2 \Leftrightarrow y_2 - y_1 \in C$, for $y_1, y_2 \in Y$. To Y we attach an abstract maximal element with respect to $\leq_C$, denoted by ∞_C and we let $Y^\bullet := Y \cup \{\infty_C\}$. Then for every $y \in Y$ one has $y \leq_C \infty_C$ and we consider on $Y^\bullet$ the following operations: $y + \infty_C = \infty_C + y = \infty_C$ and $t\infty_C = \infty_C$ for all $y \in Y$ and all $t \geq 0$. Moreover, if $\lambda \in C^*$ let $\langle \lambda, \infty_C\rangle := +\infty$.

A function $g : Y \to \overline{\mathbb{R}}$ is called *C-increasing* on a subset S of Y if for every $s_1, s_2 \in S$ such that $s_1 \leq_C s_2$ one has $g(s_1) \leq g(s_2)$.

Some of the above notions given for functions with extended real values can be formulated also for functions having their ranges in infinite-dimensional spaces.

For a function $h : X \to Y^\bullet$ we denote by $\operatorname{dom} h = \{x \in X : h(x) \in Y\}$ its *domain* and by $\operatorname{epi}_C h = \{(x, y) \in X \times Y : h(x) \leq_C y\}$ its *C-epigraph*. We say that h is *proper* if its domain is a non-empty set. The function h is said to be *C-convex* if $h(tx_1 + (1 - t)x_2) \leq_C th(x_1) + (1 - t)h(x_2)$ for all $x_1, x_2 \in X$ and all $t \in [0, 1]$. One can prove that h is C-convex if and only if $\operatorname{epi}_C h$ is a convex subset of $X \times Y$. Further, for an arbitrary $\lambda \in C^*$ we define the function $(\lambda h) : X \to \overline{\mathbb{R}}$, by $(\lambda h)(x) = \langle \lambda, h(x)\rangle$ for all $x \in X$. The function h is said to be *C-epi-closed* if $\operatorname{epi}_C h$ is a closed subset of $X \times Y$ (cf. [92]), while h is called *star C-lower semicontinuous* at $x \in X$ if for all $\lambda \in C^*$ the function (λh) is lower semicontinuous at x. The function h is said to be *star C-lower semicontinuous* if it is star C-lower semicontinuous at every $x \in X$. This notion was considered first in [84].

Remark 1.1 (a) Besides the two generalizations of lower semicontinuity defined above for functions taking values in infinite-dimensional spaces, there exist in the literature other notions of lower semicontinuity, for instance the so-called *C-lower semicontinuity*, which has been introduced by PENOT AND THÉRA in [111] and then refined in [57]. One can show that (in case $h : X \to Y$) C-lower semicontinuity implies star C-lower semicontinuity, which yields C-epi-closedness (see [92]), while the opposite assertions are not valid in general. An example of a C-convex function which is C-epi-closed, but not star C-lower semicontinuous is given in [34, Example 1]. For more on the lower semicontinuity for functions with values in topological vector spaces we refer the reader to [1, 57, 92, 111, 135].

(b) It is known that in case $Y = \mathbb{R}$ and $C = \mathbb{R}_+$ all the lower semicontinuity notions mentioned above coincide, becoming the classical lower semicontinuity of functions with real values.

Chapter 2

Regularity conditions via quasi interior and quasi-relative interior in convex optimization

Usually there is a so-called duality gap between the optimal objective values of a primal convex optimization problem and its dual problem. A challenge when dealing with duality is to give sufficient conditions which guarantee strong duality, the situation when the optimal objective values of the two problems are equal and the dual problem has an optimal solution. Several generalized interior-point regularity conditions were introduced in the literature in order to eliminate the above mentioned duality gap. Along the classical interior, some generalized interiority notions were used, like the *algebraic interior* (see [123]), the *relative algebraic interior* (see [78]), or the *strong quasi-relative interior* (see [13]), in order to state regularity conditions for strong duality. For an overview on these conditions we invite the reader to consult [71, 148] (see also [149] for more on this subject).

Nevertheless, in many theoretical and practical infinite-dimensional convex optimization problems, the interior-point conditions are useless since for instance, the interior of the set involved in the regularity condition is empty. This is the case, for example, when dealing with the positive cones ℓ_+^p and $L_+^p(T, \mu)$ of the spaces ℓ^p and $L^p(T, \mu)$, respectively, where (T, μ) is a σ-finite measure space and $p \in [1, \infty)$. For these two cones even the relative algebraic interior (which is the most general interiority notion from the aforementioned ones) is empty. In order to overcome such a situation BORWEIN AND LEWIS introduced in [14] the notion of *quasi-relative interior* of a convex set, which is a further generalization of the above mentioned interiority notions. They also proved that the quasi-relative interiors of ℓ_+^p and $L_+^p(T, \mu)$ are non-empty.

In the first section of this chapter we recall the basic properties of the above mentioned generalized interiority notions, together with the *quasi interior*, another interiority notion closely related to that of quasi-relative interior. In the following two sections we introduce several regularity conditions formulated by means of the quasi interior and quasi-relative interior which guarantee Fenchel and, respectively, Lagrange duality. The main results of this chapter are Theorem 2.3 (a Fenchel duality result) and Theorem 2.6 (a Lagrange duality result). Several examples illustrate the theoretical considerations and we provide also some comments on other regularity conditions given in the literature via the quasi-relative interior.

The theory presented in this chapter is mainly based on [26, 30].

2.1   Generalized interiority notions

We start with an overview on the most important generalized interiority notions introduced in the literature. Consider X a separated locally convex space and $U \subseteq X$ a non-empty set. The *algebraic interior* (the *core*) of U is the set (cf. [78, 123, 149])

$$\mathrm{core}(U) = \{u \in X \mid \forall x \in X, \ \exists \delta > 0 \text{ such that } \forall \lambda \in [0, \delta] : u + \lambda x \in U\},$$

while its *relative algebraic interior* (sometimes called also *intrinsic core*) is the set (cf. [78, 149])

$$\mathrm{icr}(U) = \{u \in X \mid \forall x \in \mathrm{aff}(U - U), \ \exists \delta > 0 \text{ such that } \forall \lambda \in [0, \delta] : u + \lambda x \in U\}.$$

We consider also the *strong quasi-relative interior* (sometimes called *intrinsic relative algebraic interior*) of U (cf. [13, 85, 147, 149]), denoted by $\mathrm{sqri}(U)$ (or ^{ic}U)

$$\mathrm{sqri}(U) = \begin{cases} \mathrm{icr}(U), & \text{if } \mathrm{aff}(U) \text{ is a closed set}, \\ \emptyset, & \text{otherwise}. \end{cases}$$

In this case U is a convex set, the above generalized interiority notions can be characterized as follows:

- $\mathrm{core}(U) = \{x \in U : \mathrm{cone}(U - x) = X\}$ (cf. [123, 149]);

- $\mathrm{icr}(U) = \{x \in U : \mathrm{cone}(U - x) \text{ is a linear subspace of } X\}$ (cf. [12, 78, 149]);

- $\mathrm{sqri}(U) = \{x \in U : \mathrm{cone}(U - x) \text{ is a closed linear subspace of } X\}$ (cf. [13, 85, 147, 149]);

- $x \in \mathrm{sqri}(U)$ if and only if $x \in \mathrm{icr}(U)$ and $\mathrm{aff}(U - x)$ is a closed linear subspace of X (cf. [71, 147, 149]).

The *quasi-relative interior* of U is the set (cf. [14])

$$\mathrm{qri}(U) = \{x \in U : \mathrm{cl}\big(\mathrm{cone}(U - x)\big) \text{ is a linear subspace of } X\}.$$

We give the following useful characterization of the quasi-relative interior of a convex set by means of the normal cone.

Proposition 2.1 *(cf. [14]) Let U be a non-empty convex subset of X and $x \in U$. Then $x \in \mathrm{qri}(U)$ if and only if $N_U(x)$ is a linear subspace of X^*.*

Next we consider another interiority notion introduced in connection with a convex set, which is close to the one of quasi-relative interior. The *quasi interior* of U is the set

$$\mathrm{qi}(U) = \{x \in U : \mathrm{cl}\big(\mathrm{cone}(U - x)\big) = X\}.$$

The following characterization of the quasi interior of a convex set was given in [61], where the authors supposed that X is a reflexive Banach space. One can prove that this property holds in a more general context, namely in every separated locally convex space.

Proposition 2.2 *Let U be a non-empty convex subset of X and $x \in U$. Then $x \in \mathrm{qi}(U)$ if and only if $N_U(x) = \{0\}$.*

Proof. Assume first that $x \in \text{qi}(U)$ and take an arbitrary element $x^* \in N_U(x)$. By the definition of the normal cone and the continuity of x^* we obtain $\langle x^*, z \rangle \leq 0$ for all $z \in \text{cl}\left(\text{cone}(U - x)\right)$. Thus $\langle x^*, z \rangle \leq 0$ for all $z \in X$, which is nothing else than $x^* = 0$.

In order to prove the opposite implication we consider an arbitrary $\bar{x} \in X$ and prove that $\bar{x} \in \text{cl}\left(\text{cone}(U - x)\right)$. Assuming the contrary, by a separation theorem (see for instance [149, Theorem 1.1.5]), we obtain that there exists $x^* \in X^* \setminus \{0\}$ and $\alpha \in \mathbb{R}$ such that

$$\langle x^*, z \rangle < \alpha < \langle x^*, \bar{x} \rangle \; \forall z \in \text{cl}\left(\text{cone}(U - x)\right).$$

Let $y \in U$ be fixed. For all $\lambda > 0$ it holds $\langle x^*, y - x \rangle < (1/\lambda)\alpha$ and this implies that $\langle x^*, y - x \rangle \leq 0$. As this inequality is true for every arbitrary $y \in U$, we obtain that $x^* \in N_U(x)$. But this leads to a contradiction and in this way the conclusion follows. □

Remark 2.1 Let us notice that for $x \in U$ the implication $x \in \text{qi}(U) \Rightarrow N_U(x) = \{0\}$ holds in the more general setting of topological vector spaces.

We have the following inclusions for a set $U \subseteq X$:

$$\text{int}(U) \subseteq \text{core}(U) \subseteq \begin{matrix} \text{sqri}(U) \subseteq \text{icr}(U) \\ \\ \text{qi}(U) \end{matrix} \subseteq \text{qri}(U) \subseteq U, \qquad (2.1)$$

in general the inclusions being strict. Let us suppose in the following that U is a convex set. In case $\text{int}(U) \neq \emptyset$, all the generalized interiority notions mentioned above coincide with $\text{int}(U)$ (cf. [14, Corollary 2.14]). Let us mention that if X is a Banach space and U is a closed set then $\text{core}(U) = \text{int}(U)$ (cf. [123]).

It follows from the definitions above that $\text{qri}(\{x\}) = \{x\}$ for all $x \in X$. Moreover, if $\text{qi}(U) \neq \emptyset$, then $\text{qi}(U) = \text{qri}(U)$. Although this property is given in [91] in the case of normed spaces, it holds also in separated locally convex spaces, as follows easily from the properties given above. For U, V two convex subsets of X such that $U \subseteq V$, we have $\text{qi}(U) \subseteq \text{qi}(V)$, a property which is no longer true for the quasi-relative interior (however it holds in case $\text{aff}(U) = \text{aff}(V)$, see [56, Proposition 1.12]). If X if finite-dimensional then $\text{qri}(U) = \text{sqri}(U) = \text{icr}(U) = \text{ri}(U)$ (cf. [14, 71]) and $\text{core}(U) = \text{qi}(U) = \text{int}(U)$ (cf. [91, 123]). We refer the reader to [12, 14, 20, 71, 78, 91, 123, 136, 149] and the references therein for more properties and examples regarding the above considered generalized interiority notions.

Example 2.1 Take an arbitrary $p \in [1, +\infty)$ and consider the Banach space $\ell^p = \ell^p(\mathbb{N})$ of real sequences $(x_n)_{n \in \mathbb{N}}$ such that $\sum_{n=1}^{\infty} |x_n|^p < +\infty$, equipped with the norm

$$\| \cdot \| : \ell^p \to \mathbb{R}, \; \|x\| = \left(\sum_{n=1}^{\infty} |x_n|^p \right)^{1/p} \text{ for all } x = (x_n)_{n \in \mathbb{N}} \in \ell^p. \text{ Then (cf. [14])}$$

$$\text{qri}(\ell^p_+) = \{(x_n)_{n \in \mathbb{N}} \in \ell^p : x_n > 0 \; \forall n \in \mathbb{N}\},$$

where $\ell^p_+ = \{(x_n)_{n \in \mathbb{N}} \in \ell^p : x_n \geq 0 \; \forall n \in \mathbb{N}\}$ is the positive cone of ℓ^p. Moreover, one can prove that

$$\text{int}(\ell^p_+) = \text{core}(\ell^p_+) = \text{sqri}(\ell^p_+) = \text{icr}(\ell^p_+) = \emptyset.$$

In the setting of separable Banach spaces every non-empty closed convex subset has a non-empty quasi-relative interior (cf. [14, Theorem 2.19], see also [12, Theorem 2.8] and [149, Proposition 1.2.9]) and every nonempty convex subset which is not contained in a hyperplane possesses a non-empty quasi interior (cf. [91]). The condition X is separable is important as the following example shows.

Example 2.2 For $p \in [1, +\infty)$ consider the Banach space

$$\ell^p(\mathbb{R}) = \{s : \mathbb{R} \to \mathbb{R} \mid \sum_{r \in \mathbb{R}} |s(r)|^p < \infty\},$$

equipped with the norm $\|\cdot\| : \ell^p(\mathbb{R}) \to \mathbb{R}$, $\|s\| = \left(\sum_{r \in \mathbb{R}} |s(r)|^p \right)^{1/p}$ for all $s \in \ell^p(\mathbb{R})$, where

$$\sum_{r \in \mathbb{R}} |s(r)|^p = \sup_{F \subseteq \mathbb{R}, F \text{ finite}} \sum_{r \in F} |s(r)|^p.$$

Note that the space $\ell^p(\mathbb{R})$ is not separable. Considering the positive cone $\ell^p_+(\mathbb{R}) = \{s \in \ell^p(\mathbb{R}) : s(r) \geq 0 \ \forall r \in \mathbb{R}\}$, we have (cf. [14, Example 3.11(iii)], see also [16, Remark 2.20]) that qri $(\ell^p_+(\mathbb{R})) = \emptyset$.

Useful properties of the quasi-relative interior are listed below. For the proof of $(i) - (viii)$ we refer to [12, 14].

Proposition 2.3 *Let us consider U and V two non-empty convex subsets of X, $x \in X$ and $\alpha \in \mathbb{R}$, $\alpha \neq 0$. Then*

(i) qri(U) + qri$(V) \subseteq$ qri$(U + V)$;

(ii) qri$(U \times V) =$ qri$(U) \times$ qri(V);

(iii) qri$(U - x) =$ qri$(U) - x$;

(iv) qri$(\alpha U) = \alpha$ qri(U);

(v) t qri$(U) + (1 - t)U \subseteq$ qri(U) $\forall t \in (0, 1]$, *hence* qri(U) *is a convex set;*

(vi) *if U is an affine set then* qri$(U) = U$;

(vii) qri $($qri$(U)) =$ qri(U).

If qri$(U) \neq \emptyset$ *then*

(viii) cl $($qri$(U)) =$ cl(U);

(ix) cl $\left(\text{cone}\left(\text{qri}(U)\right)\right) = $ cl $\left(\text{cone}(U)\right)$.

Proof. (ix) The inclusion cl $\left(\text{cone}\left(\text{qri}(U)\right)\right) \subseteq$ cl $\left(\text{cone}(U)\right)$ is obvious. We prove that cone$(U) \subseteq$ cl $\left(\text{cone}\left(\text{qri}(U)\right)\right)$. Consider $x \in$ cone(U) arbitrary. There exist $\lambda \geq 0$ and $u \in U$ such that $x = \lambda u$. Take $x_0 \in$ qri(U). Applying the property (v) we get $tx_0 + (1 - t)u \in$ qri(U) $\forall t \in (0, 1]$, so $\lambda t x_0 + (1 - t)x = \lambda(t x_0 + (1 - t)u) \in$ cone $($qri$(U))$ $\forall t \in (0, 1]$. Passing to the limit as $t \searrow 0$ we obtain $x \in$ cl $\left(\text{cone}\left(\text{qri}(U)\right)\right)$ and hence the desired conclusion follows. $\square$

Remark 2.2 In case $\alpha = 0$ and qri$(U) = \emptyset$, property (iv) in Proposition 2.3 above does not hold. However, when qri$(U) \neq \emptyset$, this property holds also for $\alpha = 0$.

The next lemma plays an important role in this chapter.

Lemma 2.1 *Let U and V be non-empty convex subsets of X and $x \in X$. Then*

(i) *if* qri$(U) \cap V \neq \emptyset$ *and* $0 \in$ qi$(U - U)$, *then* $0 \in$ qi$(U - V)$;

(ii) $x \in$ qi(U) *if and only if* $x \in$ qri(U) *and* $0 \in$ qi$(U - U)$.

Proof. (i) Take $y \in \text{qri}(U) \cap V$ and an arbitrary $x^* \in N_{U-V}(0)$. We get $\langle x^*, u-v \rangle \leq 0$ for all $u \in U$ and all $v \in V$. This implies

$$\langle x^*, u - y \rangle \leq 0 \ \forall u \in U, \tag{2.2}$$

that is $x^* \in N_U(y)$. As $y \in \text{qri}(U)$, $N_U(y)$ is a linear subspace of X^* (cf. Proposition 2.1), hence $-x^* \in N_U(y)$, which is nothing else than

$$\langle x^*, y - u \rangle \leq 0 \ \forall u \in U. \tag{2.3}$$

The relations (2.2) and (2.3) give us $\langle x^*, u_1 - u_2 \rangle \leq 0$ for all $u_1, u_2 \in U$, so $x^* \in N_{U-U}(0)$. Since $0 \in \text{qi}(U - U)$ we have $N_{U-U}(0) = \{0\}$ (cf. Proposition 2.2) and we get $x^* = 0$. As x^* was arbitrarily chosen we obtain $N_{U-V}(0) = \{0\}$ and, using again Proposition 2.2, the conclusion follows.

(ii) Suppose that $x \in \text{qi}(U)$. Then $x \in \text{qri}(U)$ and since $U - x \subseteq U - U$ and $0 \in \text{qi}(U-x)$, the direct implication follows. The reverse one is a direct consequence of (i) by taking $V := \{x\}$. □

Remark 2.3 Let us notice that the assertion (ii) in the above lemma can be proved directly, not necessarily by using (i).

Remark 2.4 Considering again the setting of Example 2.1 we get from the second part of the previous lemma (since $\ell_+^p - \ell_+^p = \ell^p$) that

$$\text{qi}(\ell_+^p) = \text{qri}(\ell_+^p) = \{(x_n)_{n \in \mathbb{N}} \in \ell^p : x_n > 0 \ \forall n \in \mathbb{N}\}.$$

Next we give useful separation theorems in terms of the notion of quasi-relative interior. They will be important in the next two sections in the proof of the strong duality results.

Theorem 2.1 *Let U be a non-empty convex subset of X and $x \in U$. If $x \notin \text{qri}(U)$, then there exists $x^* \in X^*, x^* \neq 0$, such that*

$$\langle x^*, y \rangle \leq \langle x^*, x \rangle \ \forall y \in U.$$

Viceversa, if there exists $x^* \in X^*$, $x^* \neq 0$, such that

$$\langle x^*, y \rangle \leq \langle x^*, x \rangle \ \forall y \in U$$

and

$$0 \in \text{qi}(U - U),$$

then $x \notin \text{qri}(U)$.

Proof. Suppose that $x \notin \text{qri}(U)$. According to Proposition 2.1, $N_U(x)$ is not a linear subspace of X^*, hence there exists $x^* \in N_U(x)$, $x^* \neq 0$. Using the definition of the normal cone, we get that $\langle x^*, y \rangle \leq \langle x^*, x \rangle$ for all $y \in U$.

Conversely, assume that there exists $x^* \in X^*$, $x^* \neq 0$, such that $\langle x^*, y \rangle \leq \langle x^*, x \rangle$ for all $y \in U$ and $0 \in \text{qi}(U - U)$. We obtain $x^* \in N_U(x)$. If we suppose that $x \in \text{qri}(U)$, then we obtain via Lemma 2.1(ii) that $x \in \text{qi}(U)$, hence (cf. Proposition 2.2) $x^* = 0$, which is a contradiction. In conclusion, $x \notin \text{qri}(U)$. □

Remark 2.5 (a) A closer look at the above proof shows that a similar separation theorem can be given in terms of the quasi interior, in which case the condition $0 \in \text{qi}(U - U)$ can be removed.

(b) Let us suppose that X is a normed space. In [60, 61] a similar separation theorem in terms of the quasi-relative interior is given. For the second part of the above theorem the authors require that the following condition must be fulfilled

$$\text{cl}\left(T_U(x) - T_U(x)\right) = X,$$

where

$$T_U(x) = \left\{ y \in X : y = \lim_{n \to \infty} \lambda_n(x_n - x), \lambda_n > 0 \ , x_n \in U \ \forall n \in \mathbb{N} \text{ and } \lim_{n \to \infty} x_n = x \right\}$$

is the so-called *contingent (Bouligand) cone* to U at $x \in U$. In general, we have the following inclusion: $T_U(x) \subseteq \mathrm{cl}\left(\mathrm{cone}(U - x)\right)$. If the set U is convex, then $T_U(x) = \mathrm{cl}\left(\mathrm{cone}(U - x)\right)$ (cf. [80]). As $\mathrm{cl}\left(\mathrm{cl}(E) + \mathrm{cl}(F)\right) = \mathrm{cl}(E + F)$, for arbitrary sets E, F in X and $\mathrm{cone}(V) - \mathrm{cone}(V) = \mathrm{cone}(V - V)$, if V is a convex subset of X such that $0 \in V$, the condition $\mathrm{cl}(T_U(x) - T_U(x)) = X$ can be reformulated as follows: $\mathrm{cl}\left(\mathrm{cone}(U - U)\right) = X$ or, equivalently, $0 \in \mathrm{qi}(U - U)$. Indeed, for $x \in U$ we have

$$\mathrm{cl}\left[\mathrm{cl}\left(\mathrm{cone}(U - x)\right) - \mathrm{cl}\left(\mathrm{cone}(U - x)\right)\right] = X \Leftrightarrow \mathrm{cl}\left[\mathrm{cone}(U - x) - \mathrm{cone}(U - x)\right] = X$$

$$\Leftrightarrow \mathrm{cl}\left(\mathrm{cone}(U - U)\right) = X \Leftrightarrow 0 \in \mathrm{qi}(U - U).$$

This means that Theorem 2.1 is a generalization to separated locally convex spaces of the separation theorem stated in [60, 61] in the framework of normed spaces.

The condition $x \in U$ in Theorem 2.1 is essential (see [61, Remark 2]). However, if x is an arbitrary element of X, one can give an alternative separation theorem based on the following result due to CAMMAROTO AND DI BELLA (cf. [55, Theorem 2.1]).

Theorem 2.2 *(cf. [55])* Let U and V be non-empty convex subsets of X with $\mathrm{qri}(U) \neq \emptyset$, $\mathrm{qri}(V) \neq \emptyset$ and such that $\mathrm{cl}\left(\mathrm{cone}\left(\mathrm{qri}(U) - \mathrm{qri}(V)\right)\right)$ is not a linear subspace of X. Then there exists $x^* \in X^*$, $x^* \neq 0$, such that $\langle x^*, u \rangle \leq \langle x^*, v \rangle$ for all $u \in U$ and all $v \in V$.

The following result is a direct consequence of Theorem 2.2.

Corollary 2.1 Let U be a non-empty convex subset of X and $x \in X$ such that $\mathrm{qri}(U) \neq \emptyset$ and $\mathrm{cl}\left(\mathrm{cone}(U - x)\right)$ is not a linear subspace of X. Then there exists $x^* \in X^*, x^* \neq 0$, such that $\langle x^*, y \rangle \leq \langle x^*, x \rangle$ for all $y \in U$.

Proof. We take in Theorem 2.2 $V := \{x\}$. Then we apply Proposition 2.3 (iii) and (ix) to obtain the conclusion.   □

Remark 2.6 Let us mention that some strict separation theorems involving the quasi-relative interior have been provided in [56].

2.2 Fenchel duality

In this section we give some new Fenchel duality results stated in terms of the quasi interior and quasi-relative interior, respectively.

Consider the convex optimization problem

$$(P_F) \quad \inf_{x \in X} \{f(x) + g(x)\},$$

where X is a separated locally convex space and $f, g : X \to \overline{\mathbb{R}}$ are proper and convex functions such that $\mathrm{dom}\, f \cap \mathrm{dom}\, g \neq \emptyset$. The Fenchel dual problem to (P_F) is

$$(D_F) \quad \sup_{x^* \in X^*} \{-f^*(-x^*) - g^*(x^*)\}.$$

We denote by $v(P_F)$ and $v(D_F)$ the optimal objective values of the primal and the dual problem, respectively. Weak duality always holds, that is $v(D_F) \leq v(P_F)$ (it follows immediately by applying the Young-Fenchel inequality). Let us recall the most important regularity conditions from the literature concerning Fenchel duality:

(RC_1^F) | $\exists x' \in \operatorname{dom} f \cap \operatorname{dom} g$ such that f (or g) is continuous at x';

(RC_2^F) | X is a Fréchet space, f and g are lower semicontinuous and $0 \in \operatorname{int}(\operatorname{dom} f - \operatorname{dom} g)$;

(RC_3^F) | X is a Fréchet space, f and g are lower semicontinuous and $0 \in \operatorname{core}(\operatorname{dom} f - \operatorname{dom} g)$;

(RC_4^F) | X is a Fréchet space, f and g are lower semicontinuous, $\operatorname{aff}(\operatorname{dom} f - \operatorname{dom} g)$ is a closed linear subspace of X and $0 \in \operatorname{icr}(\operatorname{dom} f - \operatorname{dom} g)$

and

(RC_5^F) | X is a Fréchet space, f and g are lower semicontinuous and $0 \in \operatorname{sqri}(\operatorname{dom} f - \operatorname{dom} g)$.

The condition (RC_3^F) was considered by ROCKAFELLAR (cf. [123]), (RC_5^F) by ATTOUCH AND BRÉZIS (cf. [2]), ZĂLINESCU (cf. [147]) and RODRIGUES (cf. [126]), while GOWDA AND TEBOULLE (cf. [71]) proved that (RC_4^F) and (RC_5^F) are equivalent. Let us notice that all these conditions guarantee strong duality. Moreover, if we suppose the additional hypotheses that the functions f and g are lower semicontinuous and X is a Fréchet space, between the above conditions we have the following relation: $(RC_1^F) \Rightarrow (RC_2^F) \Rightarrow (RC_3^F) \Rightarrow (RC_4^F) \Leftrightarrow (RC_5^F)$ (cf. [71], see also [149, Theorem 2.8.7]).

Remark 2.7 Let us notice that the regularity conditions (RC_2^F) and (RC_3^F) are equivalent. Indeed, assume that X is a Fréchet space, f, g are proper, convex and lower semicontinuous functions such that $\operatorname{dom} f \cap \operatorname{dom} g \neq \emptyset$ and consider the *infimal value function* $h : X \to \overline{\mathbb{R}}$, defined by $h(y) = \inf_{x \in X}\{f(x) + g(x - y)\}$ for all $y \in X$. The function h is convex and not necessarily lower semicontinuous, while one has that $\operatorname{dom} h = \operatorname{dom} f - \operatorname{dom} g$. Nevertheless, the function $(x, y) \mapsto f(x) + g(x - y)$ is ideally convex (being convex and lower semicontinuous), hence h is li-convex (cf. [149, Proposition 2.2.18]). Now by [149, Theorem 2.2.20] it follows that $\operatorname{core}(\operatorname{dom} h) = \operatorname{int}(\operatorname{dom} h)$, which has as consequence the equivalence of the regularity conditions (RC_2^F) and (RC_3^F). Let us mention that this fact has been noticed in the setting of Banach spaces by S. SIMONS in [130, Corollary 14.3].

Taking into account the relations that exist between the generalized interiority notions presented in the first section of this chapter a natural question arises: is the condition $0 \in \operatorname{qri}(\operatorname{dom} f - \operatorname{dom} g)$ sufficient for strong duality? The following example (which can be found in [71]) shows that even if we impose a stronger condition, namely $0 \in \operatorname{qi}(\operatorname{dom} f - \operatorname{dom} g)$, the above question has a negative answer and this means that we need to look for additional assumptions in order to guarantee Fenchel duality.

Example 2.3 Consider the Hilbert space $X = \ell^2(\mathbb{N})$ and the sets

$$C = \{(x_n)_{n \in \mathbb{N}} \in \ell^2 : x_{2n-1} + x_{2n} = 0 \ \forall n \in \mathbb{N}\}$$

and

$$S = \{(x_n)_{n \in \mathbb{N}} \in \ell^2 : x_{2n} + x_{2n+1} = 0 \ \forall n \in \mathbb{N}\},$$

which are closed linear subspaces of ℓ^2 and satisfy $C \cap S = \{0\}$. Define the functions $f, g : \ell^2 \to \overline{\mathbb{R}}$ by $f = \delta_C$ and $g(x) = x_1 + \delta_S(x)$, respectively, for all $x = (x_n)_{n \in \mathbb{N}} \in \ell^2$. One can see that f and g are proper, convex and lower semicontinuous functions with $\operatorname{dom} f = C$ and $\operatorname{dom} g = S$. As $v(P_F) = 0$ and $v(D_F) = -\infty$ (cf. [71,

Example 3.3]), there is a duality gap between the optimal objective values of the primal problem and its Fenchel dual problem. Moreover, $S - C$ is dense in ℓ^2 (cf. [71]), thus $\text{cl}\left(\text{cone}(\text{dom } f - \text{dom } g)\right) = \text{cl}(C - S) = \ell^2$. The last relation implies $0 \in \text{qi}(\text{dom } f - \text{dom } g)$, hence $0 \in \text{qri}(\text{dom } f - \text{dom } g)$.

Let us notice that if $v(P_F) = -\infty$, by the weak duality result follows that for the primal-dual pair $(P_F) - (D_F)$ strong duality holds. This is the reason why we suppose in the following that $v(P_F) \in \mathbb{R}$.

Lemma 2.2 *The following relation is always true*

$$0 \in \text{qri}(\text{dom } f - \text{dom } g) \Rightarrow (0,1) \in \text{qri}\left(\text{epi } f - \widehat{\text{epi}}(g - v(P_F))\right).$$

Proof. One can see that $\widehat{\text{epi}}(g - v(P_F)) = \{(x,r) \in X \times \mathbb{R} : r \leq -g(x) + v(P_F)\}$. Let us prove first that $(0,1) \in \text{epi } f - \widehat{\text{epi}}(g - v(P_F))$. Since $\inf_{x \in X}\{f(x) + g(x)\} = v(P_F) < v(P_F) + 1$, there exists $x' \in X$ such that $f(x') + g(x') < v(P_F) + 1$. Then $(0,1) = (x', v(P_F) + 1 - g(x')) - (x', -g(x') + v(P_F)) \in \text{epi } f - \widehat{\text{epi}}(g - v(P_F))$.

Now let $(x^*, r^*) \in N_{\text{epi } f - \widehat{\text{epi}}(g-v(P_F))}(0,1)$. We have

$$\langle x^*, x - x' \rangle + r^*(\mu - \mu' - 1) \leq 0 \ \forall(x,\mu) \in \text{epi } f \ \forall(x',\mu') \in \widehat{\text{epi}}(g - v(P_F)). \quad (2.\ 4)$$

For $(x,\mu) := (x_0, f(x_0))$ and $(x',\mu') := (x_0, -g(x_0) + v(P_F) - 2)$ in (2. 4), where $x_0 \in \text{dom } f \cap \text{dom } g$ is fixed, we get $r^*(f(x_0) + g(x_0) - v(P_F) + 1) \leq 0$, hence $r^* \leq 0$. As $\inf_{x \in X}\{f(x) + g(x)\} = v(P_F) < v(P_F) + 1/2$, there exists $x_1 \in X$ such that $f(x_1) + g(x_1) < v(P_F) + 1/2$. Taking now $(x,\mu) := (x_1, f(x_1))$ and $(x',\mu') := (x_1, -g(x_1) + v(P_F) - 1/2)$ in (2. 4) we obtain $r^*(f(x_1) + g(x_1) - v(P_F) - 1/2) \leq 0$ and so $r^* \geq 0$. Thus $r^* = 0$ and (2. 4) gives: $\langle x^*, x - x' \rangle \leq 0$ for all $x \in \text{dom } f$ and all $x' \in \text{dom } g$. Hence $x^* \in N_{\text{dom } f - \text{dom } g}(0)$. Since $N_{\text{dom } f - \text{dom } g}(0)$ is a linear subspace of X^* (cf. Proposition 2.1), we have $\langle -x^*, x - x' \rangle \leq 0$ for all $x \in \text{dom } f$ and $x' \in \text{dom } g$ and so $-(x^*, r^*) = (-x^*, 0) \in N_{\text{epi } f - \widehat{\text{epi}}(g-v(P_F))}(0,1)$, showing that $N_{\text{epi } f - \widehat{\text{epi}}(g-v(P_F))}(0,1)$ is a linear subspace of $X^* \times \mathbb{R}$. Hence, applying again Proposition 2.1, we get $(0,1) \in \text{qri}\left(\text{epi } f - \widehat{\text{epi}}(g - v(P_F))\right)$. $\quad\square$

Proposition 2.4 *Assume that* $0 \in \text{qi}\left[(\text{dom } f - \text{dom } g) - (\text{dom } f - \text{dom } g)\right]$. *Then* $N_{\text{co}\left[(\text{epi } f - \widehat{\text{epi}}(g-v(P_F)))\cup\{(0,0)\}\right]}(0,0)$ *is a linear subspace of* $X^* \times \mathbb{R}$ *if and only if* $N_{\text{co}\left[(\text{epi } f - \widehat{\text{epi}}(g-v(P_F)))\cup\{(0,0)\}\right]}(0,0) = \{(0,0)\}$.

Proof. The sufficiency is trivial and holds without the additional assumption from the hypotheses. Now let us suppose that $N_{\text{co}\left[(\text{epi } f - \widehat{\text{epi}}(g-v(P_F)))\cup\{(0,0)\}\right]}(0,0)$ is a linear subspace of $X^* \times \mathbb{R}$. Take $(x^*, r^*) \in N_{\text{co}\left[(\text{epi } f - \widehat{\text{epi}}(g-v(P_F)))\cup\{(0,0)\}\right]}(0,0)$. Then

$$\langle x^*, x - x' \rangle + r^*(\mu - \mu') \leq 0 \ \forall(x,\mu) \in \text{epi } f \ \forall(x',\mu') \in \widehat{\text{epi}}(g - v(P_F)). \quad (2.\ 5)$$

Let $x_0 \in \text{dom } f \cap \text{dom } g$ be fixed. Taking $(x,\mu) := (x_0, f(x_0)) \in \text{epi } f$ and $(x',\mu') := (x_0, -g(x_0) + v(P_F) - 1/2) \in \widehat{\text{epi}}(g - v(P_F))$ in the previous inequality we get $r^*(f(x_0) + g(x_0) - v(P_F) + 1/2) \leq 0$, which implies $r^* \leq 0$. Taking into account that the set $N_{\text{co}\left[(\text{epi } f - \widehat{\text{epi}}(g-v(P_F)))\cup\{(0,0)\}\right]}(0,0)$ is a linear subspace of $X^* \times \mathbb{R}$, the same argument applies also for $(-x^*, -r^*)$, implying $-r^* \leq 0$. In this way we get $r^* = 0$. From (2. 5) and the relation $(-x^*, 0) \in N_{\text{co}\left[(\text{epi } f - \widehat{\text{epi}}(g-v(P_F)))\cup\{(0,0)\}\right]}(0,0)$ we obtain

$$\langle x^*, x - x' \rangle = 0 \ \forall(x,\mu) \in \text{epi } f \ \forall(x',\mu') \in \widehat{\text{epi}}(g - v(P_F)),$$

which is nothing else than $\langle x^*, x - x' \rangle = 0$ for all $x \in \operatorname{dom} f$ and all $x' \in \operatorname{dom} g$, thus $\langle x^*, x \rangle = 0$ for all $x \in \operatorname{dom} f - \operatorname{dom} g$. Since x^* is linear and continuous, the last relation implies $\langle x^*, x \rangle = 0$ for all $x \in \operatorname{cl}\left(\operatorname{cone}\left((\operatorname{dom} f - \operatorname{dom} g) - (\operatorname{dom} f - \operatorname{dom} g)\right)\right) = X$, hence $x^* = 0$ and the conclusion follows. $\qquad\square$

Remark 2.8 (a) By (1. 1) one can see that $\operatorname{cl}\left(\operatorname{cone}\left(\operatorname{epi} f - \widehat{\operatorname{epi}}(g - v(P_F))\right)\right) = \operatorname{cl}\left[\operatorname{coneco}\left((\operatorname{epi} f - \widehat{\operatorname{epi}}(g - v(P_F))) \cup \{(0,0)\}\right)\right]$. As a consequence one has the following sequence of equivalences: $N_{\operatorname{co}\left[(\operatorname{epi} f - \widehat{\operatorname{epi}}(g - v(P_F))) \cup \{(0,0)\}\right]}(0,0)$ is a linear subspace of $X^* \times \mathbb{R} \Leftrightarrow (0,0) \in \operatorname{qri}\left[\operatorname{co}\left((\operatorname{epi} f - \widehat{\operatorname{epi}}(g - v(P_F))) \cup \{(0,0)\}\right)\right] \Leftrightarrow \operatorname{cl}\left[\operatorname{coneco}\left((\operatorname{epi} f - \widehat{\operatorname{epi}}(g - v(P_F))) \cup \{(0,0)\}\right)\right]$ is a linear subspace of $X \times \mathbb{R} \Leftrightarrow \operatorname{cl}\left(\operatorname{cone}\left(\operatorname{epi} f - \widehat{\operatorname{epi}}(g - v(P_F))\right)\right)$ is a linear subspace of $X \times \mathbb{R}$. By using Proposition 2.2, the relation $N_{\operatorname{co}\left[(\operatorname{epi} f - \widehat{\operatorname{epi}}(g - v(P_F))) \cup \{(0,0)\}\right]}(0,0) = \{(0,0)\}$ is equivalent to $(0,0) \in \operatorname{qi}\left[\operatorname{co}\left((\operatorname{epi} f - \widehat{\operatorname{epi}}(g - v(P_F))) \cup \{(0,0)\}\right)\right]$. All together, in case $0 \in \operatorname{qi}\left[(\operatorname{dom} f - \operatorname{dom} g) - (\operatorname{dom} f - \operatorname{dom} g)\right]$, the conclusion of the previous proposition can be reformulated as follows

$$\operatorname{cl}\left[\operatorname{cone}\left(\operatorname{epi} f - \widehat{\operatorname{epi}}(g - v(P_F))\right)\right] \text{ is a linear subspace of } X \times \mathbb{R}$$

$$\Leftrightarrow (0,0) \in \operatorname{qi}\left[\operatorname{co}\left((\operatorname{epi} f - \widehat{\operatorname{epi}}(g - v(P_F))) \cup \{(0,0)\}\right)\right]$$

or, equivalently,

$$(0,0) \in \operatorname{qri}\left[\operatorname{co}\left((\operatorname{epi} f - \widehat{\operatorname{epi}}(g - v(P_F))) \cup \{(0,0)\}\right)\right]$$

$$\Leftrightarrow (0,0) \in \operatorname{qi}\left[\operatorname{co}\left((\operatorname{epi} f - \widehat{\operatorname{epi}}(g - v(P_F))) \cup \{(0,0)\}\right)\right].$$

(b) One can prove that the primal problem (P_F) has an optimal solution if and only if $(0,0) \in \operatorname{epi} f - \widehat{\operatorname{epi}}(g - v(P_F))$. This means that if we suppose that the primal problem has an optimal solution and $0 \in \operatorname{qi}\left[(\operatorname{dom} f - \operatorname{dom} g) - (\operatorname{dom} f - \operatorname{dom} g)\right]$, then the conclusion of the previous proposition can be rewritten as follows: $N_{\operatorname{epi} f - \widehat{\operatorname{epi}}(g - v(P_F))}(0,0)$ is a linear subspace of $X^* \times \mathbb{R}$ if and only if we have $N_{\operatorname{epi} f - \widehat{\operatorname{epi}}(g - v(P_F))}(0,0) = \{(0,0)\}$ or, equivalently,

$$(0,0) \in \operatorname{qri}\left(\operatorname{epi} f - \widehat{\operatorname{epi}}(g - v(P_F))\right) \Leftrightarrow (0,0) \in \operatorname{qi}\left(\operatorname{epi} f - \widehat{\operatorname{epi}}(g - v(P_F))\right).$$

We introduce in the following some regularity conditions expressed in terms of the quasi interior and quasi-relative interior:

$$(RC_6^F) \quad \left| \begin{array}{l} \operatorname{dom} f \cap \operatorname{qri}(\operatorname{dom} g) \neq \emptyset, \ 0 \in \operatorname{qi}(\operatorname{dom} g - \operatorname{dom} g) \text{ and} \\ (0,0) \notin \operatorname{qri}\left[\operatorname{co}\left((\operatorname{epi} f - \widehat{\operatorname{epi}}(g - v(P_F))) \cup \{(0,0)\}\right)\right]; \end{array} \right.$$

$$(RC_7^F) \quad \left| \begin{array}{l} 0 \in \operatorname{qi}(\operatorname{dom} f - \operatorname{dom} g) \text{ and} \\ (0,0) \notin \operatorname{qri}\left[\operatorname{co}\left((\operatorname{epi} f - \widehat{\operatorname{epi}}(g - v(P_F))) \cup \{(0,0)\}\right)\right] \end{array} \right.$$

and

$$(RC_8^F) \quad \left| \begin{array}{l} 0 \in \operatorname{qi}\left[(\operatorname{dom} f - \operatorname{dom} g) - (\operatorname{dom} f - \operatorname{dom} g)\right], \\ 0 \in \operatorname{qri}(\operatorname{dom} f - \operatorname{dom} g) \text{ and} \\ (0,0) \notin \operatorname{qri}\left[\operatorname{co}\left((\operatorname{epi} f - \widehat{\operatorname{epi}}(g - v(P_F))) \cup \{(0,0)\}\right)\right]. \end{array} \right.$$

Let us prove some relations between the above considered regularity conditions.

Lemma 2.3 *Under the hypotheses we work with the following statements hold:*

(i) $(RC_6^F) \Rightarrow (RC_7^F) \Leftrightarrow (RC_8^F)$;

(ii) *in case the primal problem has an optimal solution, the condition* $(0,0) \notin$ qri $\left[\text{co}\left(\left(\text{epi}\, f - \widehat{\text{epi}}(g - v(P_F)) \right) \cup \{(0,0)\} \right) \right]$ *can be equivalently written as* $(0,0) \notin \text{qri}\left(\text{epi}\, f - \widehat{\text{epi}}(g - v(P_F)) \right)$;

(iii) *if the condition* $0 \in \text{qi}\left[(\text{dom}\, f - \text{dom}\, g) - (\text{dom}\, f - \text{dom}\, g) \right]$ *is fulfilled, then* $(0,0) \notin \text{qri}\left[\text{co}\left(\left(\text{epi}\, f - \widehat{\text{epi}}(g - v(P_F)) \right) \cup \{(0,0)\} \right) \right]$ *is equivalent to* $(0,0) \notin$ qi $\left[\text{co}\left(\left(\text{epi}\, f - \widehat{\text{epi}}(g - v(P_F)) \right) \cup \{(0,0)\} \right) \right]$.

Proof. (i) Let us suppose that (RC_6^F) is fulfilled. We apply Lemma 2.1(i) with $U := \text{dom}\, g$ and $V := \text{dom}\, f$. We get $0 \in \text{qi}(\text{dom}\, g - \text{dom}\, f)$ or, equivalently, $0 \in \text{qi}(\text{dom}\, f - \text{dom}\, g)$, that is (RC_7^F) holds. That (RC_7^F) is equivalent to (RC_8^F) is a direct consequence of Lemma 2.1(ii).

(ii) See the comment made at the beginning of Remark 2.8(b).

(iii) See Remark 2.8(a).   □

Remark 2.9 Let us notice that a sufficient condition for the fulfillment of $0 \in$ qi $\left[(\text{dom}\, f - \text{dom}\, g) - (\text{dom}\, f - \text{dom}\, g) \right]$ in Lemma 2.3(iii) is the relation $0 \in$ qi$(\text{dom}\, f - \text{dom}\, g)$. This is a direct consequence of the inclusion $\text{dom}\, f - \text{dom}\, g \subseteq (\text{dom}\, f - \text{dom}\, g) - (\text{dom}\, f - \text{dom}\, g)$.

We give now a strong duality result for the primal-dual pair $(P_F) - (D_F)$. We emphasize that for the functions f and g we suppose only convexity properties and no lower semicontinuity assumptions are needed for the duality result given below.

Theorem 2.3 *Suppose that one of the regularity conditions* (RC_i^F), $i \in \{6,7,8\}$, *is fulfilled. Then* $v(P_F) = v(D_F)$ *and* (D_F) *has an optimal solution.*

Proof. In view of Lemma 2.3(i), it is enough to give the proof in case (RC_8^F) is fulfilled, a condition which we assume in the following to be true.

Lemma 2.2 ensures that $(0,1) \in \text{qri}\left(\text{epi}\, f - \widehat{\text{epi}}(g - v(P_F)) \right)$, hence qri $\left(\text{epi}\, f - \widehat{\text{epi}}(g - v(P_F)) \right) \neq \emptyset$. The condition $(0,0) \notin \text{qri}\left[\text{co}\left(\left(\text{epi}\, f - \widehat{\text{epi}}(g - v(P_F)) \right) \cup \{(0,0)\} \right) \right]$, together with the relation cl $\left[\text{coneco}\left(\left(\text{epi}\, f - \widehat{\text{epi}}(g - v(P_F)) \right) \cup \{(0,0)\} \right) \right]$ $= \text{cl}\left(\text{cone}\left(\text{epi}\, f - \widehat{\text{epi}}(g - v(P_F)) \right) \right)$ (cf. (1. 1)), imply that cl $\left(\text{cone}\left(\text{epi}\, f - \widehat{\text{epi}}(g - v(P_F)) \right) \right)$ is not a linear subspace of $X \times \mathbb{R}$. We apply Corollary 2.1 with $U := \text{epi}\, f - \widehat{\text{epi}}(g - v(P_F))$ and $x = (0,0)$. Thus there exists $(x^*, \lambda) \in X^* \times \mathbb{R}$, $(x^*, \lambda) \neq (0,0)$, such that

$$\langle x^*, x \rangle + \lambda\mu \geq \langle x^*, x' \rangle + \lambda\mu' \ \forall (x,\mu) \in \widehat{\text{epi}}(g - v(P_F)) \ \forall (x',\mu') \in \text{epi}\, f. \quad (2.6)$$

We claim that $\lambda \leq 0$. Indeed, if $\lambda > 0$, then for $(x,\mu) := (x_0, -g(x_0) + v(P_F))$ and $(x',\mu') := (x_0, f(x_0) + n), n \in \mathbb{N}$, where $x_0 \in \text{dom}\, f \cap \text{dom}\, g$ is fixed, we obtain from (2. 6) that $\langle x^*, x_0 \rangle + \lambda(-g(x_0) + v(P_F)) \geq \langle x^*, x_0 \rangle + \lambda(f(x_0) + n)$ for all $n \in \mathbb{N}$. Passing to the limit as $n \to +\infty$ we obtain a contradiction. Next we prove that $\lambda < 0$. Suppose that $\lambda = 0$. Then from (2. 6) we have $\langle x^*, x \rangle \geq \langle x^*, x' \rangle$ for all $x \in \text{dom}\, g$ and $x' \in \text{dom}\, f$, hence $\langle x^*, x \rangle \leq 0$ for all $x \in \text{dom}\, f - \text{dom}\, g$. Using the condition $0 \in \text{qi}\left[(\text{dom}\, f - \text{dom}\, g) - (\text{dom}\, f - \text{dom}\, g) \right]$ and the second part of

Theorem 2.1 we obtain $0 \notin \mathrm{qri}(\mathrm{dom}\, f - \mathrm{dom}\, g)$, which contradicts the condition $0 \in \mathrm{qri}(\mathrm{dom}\, f - \mathrm{dom}\, g)$ from (RC_8^F). Thus we must have $\lambda < 0$ and from (2. 6) we obtain:

$$\left\langle \frac{1}{\lambda}x^*, x \right\rangle + \mu \le \left\langle \frac{1}{\lambda}x^*, x' \right\rangle + \mu' \ \forall (x,\mu) \in \widehat{\mathrm{epi}}(g - v(P_F)) \ \forall (x',\mu') \in \mathrm{epi}\, f.$$

Let $r \in \mathbb{R}$ be such that

$$\mu' + \langle x_0^*, x' \rangle \ge r \ge \mu + \langle x_0^*, x \rangle \ \forall (x,\mu) \in \widehat{\mathrm{epi}}(g - v(P_F)) \ \forall (x',\mu') \in \mathrm{epi}\, f,$$

where $x_0^* := (1/\lambda)x^*$. The first inequality yields $f(x) \ge \langle -x_0^*, x \rangle + r$ for all $x \in X$, that is $f^*(-x_0^*) \le -r$. The second one gives us $-g(x) + v(P_F) + \langle x_0^*, x \rangle \le r$ for all $x \in X$, hence $g^*(x_0^*) \le r - v(P_F)$ and so we have $-f^*(-x_0^*) - g^*(x_0^*) \ge r + v(P_F) - r = v(P_F)$. This implies that $v(D_F) \ge v(P_F)$. As the opposite inequality is always true, we get $v(P_F) = v(D_F)$ and x_0^* is an optimal solution of the problem (D_F). $\qquad\square$

Remark 2.10 (a) The proof given above relies on the separation result given in Corollary 2.1. Let us notice that alternatively, one can apply Theorem 2.1 with $U := \mathrm{co}\left(\left(\mathrm{epi}\, f - \widehat{\mathrm{epi}}(g - v(P_F))\right) \cup \{(0,0)\}\right)$ and $x := (0,0) \in U$. Relation (2. 6) follows and the proof can be continued as above.

(b) If the condition $(0,0) \notin \mathrm{qri}\left[\mathrm{co}\left(\left(\mathrm{epi}\, f - \widehat{\mathrm{epi}}(g - v(P_F))\right) \cup \{(0,0)\}\right)\right]$ is removed, the duality result given above may fail. By using again Example 2.3 we show that this condition is essential. Let us notice that for the problem in Example 2.3 the condition $0 \in \mathrm{qi}(\mathrm{dom}\, f - \mathrm{dom}\, g)$ is fulfilled and 0 is the unique optimal solution of the primal problem. We prove in the following that in the aforementioned example we have $(0,0) \in \mathrm{qri}\left(\mathrm{epi}\, f - \widehat{\mathrm{epi}}(g - v(P_F))\right)$. Note that the scalar product on ℓ^2, $\langle \cdot, \cdot \rangle : \ell^2 \times \ell^2 \to \mathbb{R}$ is given by $\langle x, y \rangle = \sum\limits_{n=1}^{\infty} x_n y_n$, for all $x = (x_n)_{n \in \mathbb{N}}, y = (y_n)_{n \in \mathbb{N}} \in \ell^2$. For $k \in \mathbb{N}$, we denote by $e^{(k)}$ the element in ℓ^2 such that $e_n^{(k)} = 1$ if $n = k$ and $e_n^{(k)} = 0$ for all $n \in \mathbb{N} \setminus \{k\}$. We have $\mathrm{epi}\, f = C \times [0, \infty)$. Further, $\widehat{\mathrm{epi}}(g - v(P_F)) = \{(x,r) \in \ell^2 \times \mathbb{R} : r \le -g(x)\} = \{(x,r) \in \ell^2 \times \mathbb{R} : x = (x_n)_{n \in \mathbb{N}} \in S, r \le -x_1\} = \{(x, -x_1 - \varepsilon) \in \ell^2 \times \mathbb{R} : x = (x_n)_{n \in \mathbb{N}} \in S, \varepsilon \ge 0\}$. Then $A := \mathrm{epi}\, f - \widehat{\mathrm{epi}}(g - v(P_F)) = \{(x - x', x_1' + \varepsilon) : x \in C, x' = (x_n')_{n \in \mathbb{N}} \in S, \varepsilon \ge 0\}$. Take $(x^*, r^*) \in N_A(0,0)$, where $x^* = (x_n^*)_{n \in \mathbb{N}} \in \ell^2$ and $r^* \in \mathbb{R}$. We have

$$\langle x^*, x - x' \rangle + r^*(x_1' + \varepsilon) \le 0 \ \forall x \in C \ \forall x' = (x_n')_{n \in \mathbb{N}} \in S \ \forall \varepsilon \ge 0. \tag{2. 7}$$

Taking in (2. 7) $x' = 0$ and $\varepsilon = 0$ we get $\langle x^*, x \rangle \le 0$ for all $x \in C$. As C is a linear subspace of X we obtain

$$\langle x^*, x \rangle = 0 \ \forall x \in C. \tag{2. 8}$$

Since $e^{(2k-1)} - e^{(2k)} \in C$ for all $k \in \mathbb{N}$, relation (2. 8) implies

$$x_{2k-1}^* - x_{2k}^* = 0 \ \forall k \in \mathbb{N}. \tag{2. 9}$$

From (2. 7) and (2. 8) we obtain

$$\langle -x^*, x' \rangle + r^*(x_1' + \varepsilon) \le 0 \ \forall x' = (x_n')_{n \in \mathbb{N}} \in S \ \forall \varepsilon \ge 0. \tag{2. 10}$$

Taking $\varepsilon = 0$ and $x' := me^1 \in S$ in (2. 10), where $m \in \mathbb{Z}$ is arbitrary, we get $m(-x_1^* + r^*) \le 0$ for all $m \in \mathbb{Z}$, thus $r^* = x_1^*$. For $\varepsilon = 0$ in (2. 10) we obtain $-\sum\limits_{n=1}^{\infty} x_n^* x_n' + r^* x_1' \le 0$ for all $x' \in S$. Taking into account that $r^* = x_1^*$, we get $-\sum\limits_{n=2}^{\infty} x_n^* x_n' \le 0$ for all $x' \in S$. As S is a linear subspace of X it follows $\sum\limits_{n=2}^{\infty} x_n^* x_n' = 0$

for all $x' \in S$, but, since $e^{(2k)} - e^{(2k+1)} \in S$ for all $k \in \mathbb{N}$, the above relation shows that

$$x_{2k}^* - x_{2k+1}^* = 0 \ \forall k \in \mathbb{N}. \tag{2.11}$$

Combining (2. 9) with (2. 11) we get $x^* = 0$ (since $x^* \in \ell^2$). Because $r^* = x_1^*$, we have also $r^* = 0$. Thus $N_A(0,0) = \{(0,0)\}$ and Proposition 2.2 leads to the desired conclusion.

(c) We have the following implication

$$(0,0) \in \mathrm{qi} \left[\mathrm{co} \left(\left(\mathrm{epi}\, f - \widehat{\mathrm{epi}}(g - v(P_F)) \right) \cup \{(0,0)\} \right) \right] \Rightarrow 0 \in \mathrm{qi}(\mathrm{dom}\, f - \mathrm{dom}\, g).$$

Indeed, suppose that $(0,0) \in \mathrm{qi} \left[\mathrm{co} \left(\left(\mathrm{epi}\, f - \widehat{\mathrm{epi}}(g - v(P_F)) \right) \cup \{(0,0)\} \right) \right]$. Then $\mathrm{cl} \left[\mathrm{coneco} \left(\left(\mathrm{epi}\, f - \widehat{\mathrm{epi}}(g - v(P_F)) \right) \cup \{(0,0)\} \right) \right] = X \times \mathbb{R}$, hence (cf. (1. 1))

$$\mathrm{cl} \left[\mathrm{cone} \left(\mathrm{epi}\, f - \widehat{\mathrm{epi}}(g - v(P_F)) \right) \right] = X \times \mathbb{R}.$$

Since the inclusion

$$\mathrm{cl} \left[\mathrm{cone} \left(\mathrm{epi}\, f - \widehat{\mathrm{epi}}(g - v(P_F)) \right) \right] \subseteq \mathrm{cl} \left(\mathrm{cone}(\mathrm{dom}\, f - \mathrm{dom}\, g) \right) \times \mathbb{R}$$

trivially holds, we have $\mathrm{cl} \left(\mathrm{cone}(\mathrm{dom}\, f - \mathrm{dom}\, g) \right) = X$, that is $0 \in \mathrm{qi}(\mathrm{dom}\, f - \mathrm{dom}\, g)$. Hence the following implication is fulfilled

$$0 \notin \mathrm{qi}(\mathrm{dom}\, f - \mathrm{dom}\, g) \Rightarrow (0,0) \notin \mathrm{qi} \left[\mathrm{co} \left(\left(\mathrm{epi}\, f - \widehat{\mathrm{epi}}(g - v(P_F)) \right) \cup \{(0,0)\} \right) \right].$$

Nevertheless, in the regularity conditions given above one cannot substitute the condition $(0,0) \notin \mathrm{qi} \left[\mathrm{co} \left(\left(\mathrm{epi}\, f - \widehat{\mathrm{epi}}(g - v(P_F)) \right) \cup \{(0,0)\} \right) \right]$ by the stronger, but more handleable one $0 \notin \mathrm{qi}(\mathrm{dom}\, f - \mathrm{dom}\, g)$, since in all the regularity conditions (RC_i^F), $i \in \{6,7,8\}$, the other hypotheses imply $0 \in \mathrm{qi}(\mathrm{dom}\, f - \mathrm{dom}\, g)$ (cf. Lemma 2.3).

Let us give in the following an example which illustrates the applicability of the strong duality result introduced above.

Example 2.4 Consider the Hilbert space $\ell^2 = \ell^2(\mathbb{N})$. We define the functions $f, g : \ell^2 \to \overline{\mathbb{R}}$ by

$$f(x) = \begin{cases} \|x\|, & \text{if } x \in x^0 - \ell_+^2, \\ +\infty, & \text{otherwise} \end{cases}$$

and

$$g(x) = \begin{cases} \langle c, x \rangle, & \text{if } x \in \ell_+^2, \\ +\infty, & \text{otherwise}, \end{cases}$$

respectively, where $x^0, c \in \ell_+^2$ are arbitrary chosen such that $x_n^0 > 0$ for all $n \in \mathbb{N}$. Note that

$$v(P_F) = \inf_{x \in \ell_+^2 \cap (x^0 - \ell_+^2)} \{\|x\| + \langle c, x \rangle\} = 0$$

and the infimum is attained at $x = 0$. We have $\mathrm{dom}\, f = x^0 - \ell_+^2 = \{(x_n)_{n \in \mathbb{N}} \in \ell^2 : x_n \leq x_n^0 \ \forall n \in \mathbb{N}\}$ and $\mathrm{dom}\, g = \ell_+^2$. By using Example 2.1 we get

$$\mathrm{dom}\, f \cap \mathrm{qri}(\mathrm{dom}\, g) = \{(x_n)_{n \in \mathbb{N}} \in \ell^2 : 0 < x_n \leq x_n^0 \ \forall n \in \mathbb{N}\} \neq \emptyset.$$

Also, $\mathrm{cl} \left(\mathrm{cone}(\mathrm{dom}\, g - \mathrm{dom}\, g) \right) = \ell^2$ and so $0 \in \mathrm{qi}(\mathrm{dom}\, g - \mathrm{dom}\, g)$. Further, $\mathrm{epi}\, f = \{(x,r) \in \ell^2 \times \mathbb{R} : x \in x^0 - \ell_+^2, \|x\| \leq r\} = \{(x, \|x\| + \varepsilon) \in \ell^2 \times \mathbb{R} : x \in x^0 - \ell_+^2, \varepsilon \geq 0\}$ and $\widehat{\mathrm{epi}}(g - v(P_F)) = \{(x,r) \in \ell^2 \times \mathbb{R} : r \leq -g(x)\} = \{(x,r) \in \ell^2 \times \mathbb{R} : r \leq$

$-\langle c, x \rangle, x \in \ell_+^2 \} = \{ (x, -\langle c, x \rangle - \varepsilon) : x \in \ell_+^2, \varepsilon \geq 0 \}$. We get epi $f - \widehat{\mathrm{epi}}(g - v(P_F)) = \{ (x - x', \|x\| + \varepsilon + \langle c, x' \rangle + \varepsilon') : x \in x^0 - \ell_+^2, x' \in \ell_+^2, \varepsilon, \varepsilon' \geq 0 \}$, hence

$$\mathrm{epi}\, f - \widehat{\mathrm{epi}}(g - v(P_F)) = \{ (x - x', \|x\| + \langle c, x' \rangle + \varepsilon) : x \in x^0 - \ell_+^2, x' \in \ell_+^2, \varepsilon \geq 0 \}.$$

In the following we prove that $(0,0) \notin \mathrm{qri}\,(\mathrm{epi}\, f - \widehat{\mathrm{epi}}(g - v(P_F)))$. Assuming the contrary, we would have that the set $\mathrm{cl}\,[\mathrm{cone}\,(\mathrm{epi}\, f - \widehat{\mathrm{epi}}(g - v(P_F)))]$ is a linear subspace of $\ell^2 \times \mathbb{R}$. Since $(0,1) \in \mathrm{cl}\,[\mathrm{cone}\,(\mathrm{epi}\, f - \widehat{\mathrm{epi}}(g - v(P_F)))]$ (take $x = x' = 0$ and $\varepsilon = 1$) we must have also that $(0,-1)$ belongs to this set. On the other hand, one can easily see that for all (x,r) belonging to $\mathrm{cl}\,[\mathrm{cone}\,(\mathrm{epi}\, f - \widehat{\mathrm{epi}}(g - v(P_F)))]$ it holds $r \geq 0$. This leads to the desired contradiction.

Hence the regularity condition (RC_6^F) is fulfilled, thus strong duality holds (cf. Theorem 2.3). On the other hand, ℓ^2 is a Fréchet space (being a Hilbert space), the functions f and g are proper, convex and lower semicontinuous and, as $\mathrm{sqri}(\mathrm{dom}\, f - \mathrm{dom}\, g) = \mathrm{sqri}(x^0 - \ell_+^2) = \emptyset$, none of the regularity conditions (RC_i^F), $i \in \{1,2,3,4,5\}$, presented at the beginning of this section can be applied for this optimization problem.

As for all $x^* \in \ell^2$ it holds $g^*(x^*) = \delta_{c-\ell_+^2}(x^*)$ and (cf. [149, Theorem 2.8.7])

$$f^*(-x^*) = \inf_{x_1^* + x_2^* = -x^*} \{ \| \cdot \|^*(x_1^*) + \delta_{x^0 - l_+^2}^*(x_2^*) \} = \inf_{\substack{x_1^* + x_2^* = -x^*, \\ \|x_1^*\| \leq 1, x_2^* \in \ell_+^2}} \langle x_2^*, x^0 \rangle,$$

the optimal objective value of the Fenchel dual problem is

$$v(D_F) = \sup_{\substack{x_2^* \in \ell_+^2 - c - x_1^*, \\ \|x_1^*\| \leq 1, x_2^* \in \ell_+^2}} \langle -x_2^*, x^0 \rangle = \sup_{x_2^* \in \ell_+^2} \langle -x_2^*, x^0 \rangle = 0,$$

while $x_2^* = 0$ is the optimal solution of the dual.

The following example underlines the fact that in general the regularity condition (RC_7^F) (and automatically also (RC_8^F), see Lemma 2.3(i)) is weaker than (RC_6^F) (see also Example 2.7 below).

Example 2.5 Consider the Hilbert space $\ell^2(\mathbb{R})$ and the functions $f, g : \ell^2(\mathbb{R}) \to \overline{\mathbb{R}}$ defined for all $s \in \ell^2(\mathbb{R})$ by

$$f(s) = \begin{cases} s(1), & \text{if } s \in \ell_+^2(\mathbb{R}), \\ +\infty, & \text{otherwise} \end{cases}$$

and

$$g(s) = \begin{cases} s(2), & \text{if } s \in \ell_+^2(\mathbb{R}), \\ +\infty, & \text{otherwise}, \end{cases}$$

respectively. The optimal objective value of the primal problem is

$$v(P_F) = \inf_{s \in \ell_+^2(\mathbb{R})} \{ s(1) + s(2) \} = 0$$

and $s = 0$ is an optimal solution (let us notice that the primal problem has infinitely many optimal solutions). We have $\mathrm{qri}(\mathrm{dom}\, g) = \mathrm{qri}(\ell_+^2(\mathbb{R})) = \emptyset$ (cf. Example 2.2), hence the condition (RC_6^F) fails. In the following we show that (RC_7^F) is fulfilled. One can prove that $\mathrm{dom}\, f - \mathrm{dom}\, g = \ell_+^2(\mathbb{R}) - \ell_+^2(\mathbb{R}) = \ell^2(\mathbb{R})$, thus $0 \in \mathrm{qi}(\mathrm{dom}\, f - \mathrm{dom}\, g)$. Like in the previous example, the following relation holds

$$\mathrm{epi}\, f - \widehat{\mathrm{epi}}(g - v(P_F)) = \{ (s - s', s(1) + s'(2) + \varepsilon) : s, s' \in \ell_+^2(\mathbb{R}), \varepsilon \geq 0 \}$$

and with the same technique one can show that $(0,0) \notin \mathrm{qri}\left(\mathrm{epi}\, f - \widehat{\mathrm{epi}}(g - v(P_F))\right)$. Thus (RC_7^F) is fulfilled and, as a consequence strong duality holds (cf. Theorem 2.3).

Let us take a look at the dual problem. For this we have to calculate the conjugates of f and g. Let us recall that the scalar product on $\ell^2(\mathbb{R})$, $\langle \cdot, \cdot \rangle : \ell^2(\mathbb{R}) \times \ell^2(\mathbb{R}) \to \mathbb{R}$ is defined by $\langle s, s' \rangle = \sup\limits_{F \subseteq \mathbb{R}, F\text{finite}} \sum_{r \in F} s(r)s'(r)$, for $s, s' \in \ell^2(\mathbb{R})$. The dual space $\left(\ell^2(\mathbb{R})\right)^*$ is identified with $\ell^2(\mathbb{R})$. For an arbitrary $u \in \ell^2(\mathbb{R})$ we have

$$f^*(u) = \sup_{s \in \ell_+^2(\mathbb{R})} \{\langle u, s \rangle - s(1)\} = \sup_{s \in \ell_+^2(\mathbb{R})} \left\{ \sup_{F \subseteq \mathbb{R}, F\text{finite}} \sum_{r \in F} u(r)s(r) - s(1) \right\}$$

$$= \sup_{F \subseteq \mathbb{R}, F\text{finite}} \left\{ \sup_{s \in \ell_+^2(\mathbb{R})} \left\{ \sum_{r \in F} u(r)s(r) - s(1) \right\} \right\}.$$

Consider $F = \{r_1, ..., r_k\}$ an arbitrary finite subset of $\mathbb{R}$, where $k \in \mathbb{N}$. The inner supremum can be written as

$$\sup_{s \in \ell_+^2(\mathbb{R})} \{u(r_1)s(r_1) + ... + u(r_k)s(r_k) - s(1)\}.$$

One can easily prove that if $1 \notin F$ this supremum is equal to 0 if $u(r_i) \leq 0$ for all $i \in \{1, ..., k\}$, being $+\infty$, otherwise. If $1 \in F$, with $r_{i_0} = 1$ ($i_0 \in \{1, ..., k\}$), the supremum becomes 0, in case $u(r_i) \leq 0$ for all $i \in \{i, ..., k\} \setminus \{i_0\}$ and $u(1) \leq 1$, being $+\infty$, otherwise. In conclusion,

$$f^*(u) = \begin{cases} 0, & \text{if } u(r) \leq 0 \ \forall r \in \mathbb{R} \setminus \{1\} \text{ and } u(1) \leq 1, \\ +\infty, & \text{otherwise.} \end{cases}$$

Similarly we compute g^* and obtain that $v(D_F) = 0$ and $u = 0$ is an optimal solution of the dual (moreover, the dual has infinitely many optimal solutions).

Let us mention that besides the above mentioned generalized interior-point regularity conditions, there exist in the literature the so-called *closedness-type regularity conditions*, considered by BURACHIK AND JEYAKUMAR in Banach spaces (cf. [47]) and by BOȚ AND WANKA in separated locally convex spaces (cf. [39]). Let us consider the following condition:

$$(RC_9^F) \quad \left| \begin{array}{l} f \text{ and } g \text{ are lower semicontinuous and} \\ \mathrm{epi}\, f^* + \mathrm{epi}\, g^* \text{ is closed in } (X^*, w(X^*, X)) \times \mathbb{R}. \end{array} \right.$$

We have the following duality result (cf. [39]).

Theorem 2.4 *Let $f, g : X \to \overline{\mathbb{R}}$ be proper and convex functions such that $\mathrm{dom}\, f \cap \mathrm{dom}\, g \neq \emptyset$. If (RC_9^F) is fulfilled, then*

$$(f + g)^*(x^*) = \min\{f^*(x^* - y^*) + g^*(y^*) : y^* \in X^*\} \ \forall x^* \in X^*. \qquad (2.12)$$

Remark 2.11 (a) Let us notice that in the literature condition (2. 12) is referred to *stable strong duality* (see [21, 48, 130] for more details) and obviously guarantees strong duality for $(P_F) - (D_F)$. When $f, g : X \to \overline{\mathbb{R}}$ are proper, convex and lower semicontinuous functions with $\mathrm{dom}\, f \cap \mathrm{dom}\, g \neq \emptyset$, the condition (RC_9^F) is fulfilled if and only if (2. 12) holds (cf. [39, Theorem 3.2]).

(b) In case X is a Fréchet space and f, g are proper, convex and lower semicontinuous functions we have the following relations between the regularity conditions considered for the primal-dual pair $(P_F)-(D_F)$ (cf. [39], see also [71] and [149, Theorem 2.8.7])

$$(RC_1^F) \Rightarrow (RC_2^F) \Leftrightarrow (RC_3^F) \Rightarrow (RC_4^F) \Leftrightarrow (RC_5^F) \Rightarrow (RC_9^F).$$

We refer to [21, 39, 47, 130] for several examples showing that in general the implications above are strict. The implication $(RC_1^F) \Rightarrow (RC_9^F)$ holds in the general setting of separated locally convex spaces (in the hypotheses that f, g are proper, convex and lower semicontinuous).

We observe that if X is a finite-dimensional space and f, g are proper, convex and lower semicontinuous, then $(RC_6^F) \Rightarrow (RC_7^F) \Leftrightarrow (RC_8^F) \Rightarrow (RC_9^F)$. However, in the infinite-dimensional setting this is no longer true. In the following two examples we show that in general the conditions (RC_7^F) (and automatically also (RC_8^F), cf. Lemma 2.3(i)) and (RC_9^F) are not comparable. In the example below, (RC_9^F) is fulfilled, unlike (RC_i^F), $i \in \{6, 7, 8\}$ (we refer to [21, 39, 47, 90, 130] for examples in the finite-dimensional setting).

Example 2.6 Consider the Hilbert space $\ell^2(\mathbb{R})$ and the functions $f, g : \ell^2(\mathbb{R}) \to \overline{\mathbb{R}}$, defined by $f = \delta_{\ell_+^2(\mathbb{R})}$ and $g = \delta_{-\ell_+^2(\mathbb{R})}$, respectively. We have qri(dom f − dom g) = qri $\left(\ell_+^2(\mathbb{R})\right) = \emptyset$ (cf. Example 2.2), hence all the generalized interior-point regularity conditions (RC_i^F), $i \in \{1, 2, 3, 4, 5, 6, 7, 8\}$ fail (cf. Remark 2.11(b) and Lemma 2.3(i)). The conjugate functions of f and g are $f^* = \delta_{-\ell_+^2(\mathbb{R})}$ and $g^* = \delta_{\ell_+^2(\mathbb{R})}$, respectively, hence epi f^* + epi g^* = $\ell^2(\mathbb{R}) \times [0, \infty)$, that is the condition (RC_9^F) holds. One can see that $v(P_F) = v(D_F) = 0$ and $y^* = 0$ is an optimal solution of the dual problem.

In the following we provide an example for which this time (RC_7^F) (and automatically also (RC_8^F), cf. Lemma 2.3(i)) is fulfilled, unlike (RC_9^F).

Example 2.7 Like in Example 2.3, consider the Hilbert space $X = \ell^2(\mathbb{N})$ and the sets

$$C = \{(x_n)_{n \in \mathbb{N}} \in \ell^2 : x_{2n-1} + x_{2n} = 0 \ \forall n \in \mathbb{N}\}$$

and

$$S = \{(x_n)_{n \in \mathbb{N}} \in \ell^2 : x_{2n} + x_{2n+1} = 0 \ \forall n \in \mathbb{N}\},$$

which are closed linear subspaces of ℓ^2 and satisfy $C \cap S = \{0\}$. Define the functions $f, g : \ell^2 \to \overline{\mathbb{R}}$ by $f = \delta_C$ and $g = \delta_S$, respectively, which are proper, convex and lower semicontinuous. The optimal objective value of the primal problem is $v(P_F) = 0$ and $\overline{x} = 0$ is the unique optimal solution of $v(P_F)$. Moreover, $S - C$ is dense in ℓ^2 (cf. [71, Example 3.3]), thus cl $\left(\text{cone}(\text{dom } f - \text{dom } g)\right) = \text{cl}(C - S) = \ell^2$. This implies $0 \in \text{qri}(\text{dom } f - \text{dom } g)$. Further, one has

$$\text{epi } f - \widehat{\text{epi}}(g - v(P_F)) = \{(x - y, \varepsilon) : x \in C, y \in S, \varepsilon \geq 0\} = (C - S) \times [0, +\infty)$$

and cl $\left[\text{cone} \left(\text{epi } f - \widehat{\text{epi}}(g - v(P_F)) \right) \right] = \ell^2 \times [0, +\infty)$, which is not a linear subspace of $\ell^2 \times \mathbb{R}$, hence $(0, 0) \notin \text{qri} \left(\text{epi } f - \widehat{\text{epi}}(g - v(P_F)) \right)$. All together, we get that the condition (RC_7^F) is fulfilled, hence strong duality holds (cf. Theorem 2.3). One can prove that $f^* = \delta_{C^\perp}$ and $g^* = \delta_{S^\perp}$, where

$$C^\perp = \{(x_n)_{n \in \mathbb{N}} \in \ell^2 : x_{2n-1} = x_{2n} \ \forall n \in \mathbb{N}\}$$

and

$$S^\perp = \{(x_n)_{n \in \mathbb{N}} \in \ell^2 : x_1 = 0, x_{2n} = x_{2n+1} \ \forall n \in \mathbb{N}\}.$$

Further, $v(D_F) = 0$ and the set of optimal solutions of the dual problem is exactly $C^\perp \cap S^\perp = \{0\}$.

We show that (RC_9^F) is not fulfilled. Let us consider the element $e^{(1)} \in \ell^2$. We compute $(f + g)^*(e^{(1)}) = \sup_{x \in \ell^2} \{\langle e^{(1)}, x \rangle - f(x) - g(x)\} = 0$ and $(f^* \Box g^*)(e^{(1)}) = \delta_{C^\perp + S^\perp}(e^{(1)})$. If we suppose that $e^{(1)} \in C^\perp + S^\perp$, then we would have $(e^{(1)} + S^\perp) \cap C^\perp \neq \emptyset$. However, it has been proved in [71, Example 3.3] that $(e^{(1)} + S^\perp) \cap C^\perp = \emptyset$. This shows that $(f^* \Box g^*)(e^{(1)}) = +\infty > 0 = (f + g)^*(e^{(1)})$. Via Theorem 2.4 follows that the condition (RC_9^F) is not fulfilled and, consequently, (RC_i^F), $i \in \{1, 2, 3, 4, 5\}$, fail, too (cf. Remark 2.11(b)), unlike condition (RC_7^F). Looking at (RC_6^F), one can see that this condition is also not fulfilled, since $0 \in \mathrm{qi}(\mathrm{dom}\, g - \mathrm{dom}\, g)$ does not hold.

Finally, let us notice that one can prove directly that (RC_9^F) is not fulfilled. Indeed, we have $\mathrm{epi}\, f^* + \mathrm{epi}\, g^* = (C^\perp + S^\perp) \times [0, \infty)$. As in [71, Example 3.3], one can show that $C^\perp + S^\perp$ is dense in ℓ^2. If we suppose that $C^\perp + S^\perp$ is closed, we would have $C^\perp + S^\perp = \ell^2$, which is a contradiction, since $e^{(1)} \notin C^\perp + S^\perp$.

Remark 2.12 Let us notice that under convexity assumptions of the functions involved, C. LI, D. FANG, G. LÓPEZ AND M.A. LÓPEZ introduced in [90] a condition which equivalently characterizes stable strong duality, that is relation (2. 12) (cf. [90, Theorem 4.6]). This condition looks like (cf. [90, Definition 3.1, Lemma 3.3 and relation (3.5)]):

$$(CQ^{LFLL}) \quad | \quad \mathrm{epi}(f + g)^* = \mathrm{epi}\, f^* + \mathrm{epi}\, g^*.$$

As noticed in [90, Corollary 3.9], in case f, g are proper, convex and lower semicontinuous functions, the conditions (CQ^{LFLL}) and (RC_9^F) are equivalent. The authors gave also an example (in the finite-dimensional setting) for which (CQ^{LFLL}) holds, but the regularity condition expressed by means of the quasi interior and quasi-relative interior (RC_8^F) fails (cf. [90, Example 4.1]). Example 2.7 above provides a situation where the condition (RC_8^F) is fulfilled, unlike (CQ^{LFLL}).

In the following, by using the results introduced above, we give regularity conditions for the following convex optimization problem

$$(P_F^A) \quad \inf_{x \in X} \{f(x) + (g \circ A)(x)\},$$

where X and Y are separated locally convex spaces having as topological dual spaces X^* and Y^*, respectively, $A : X \to Y$ is a continuous linear mapping, $f : X \to \overline{\mathbb{R}}$ and $g : Y \to \overline{\mathbb{R}}$ are proper and convex functions such that $A(\mathrm{dom}\, f) \cap \mathrm{dom}\, g \neq \emptyset$. The Fenchel dual problem to (P_F^A) is

$$(D_F^A) \quad \sup_{y^* \in Y^*} \{-f^*(-A^*y^*) - g^*(y^*)\}.$$

We denote by $v(P_F^A)$ and $v(D_F^A)$ the optimal objective values of the primal and the dual problem, respectively. We suppose also that $v(P_F^A) \in \mathbb{R}$. We consider the set

$$A \times \mathrm{id}_{\mathbb{R}}(\mathrm{epi}\, f) = \{(Ax, r) \in Y \times \mathbb{R} : f(x) \leq r\}.$$

Let us introduce the following functions: $F, G : X \times Y \to \overline{\mathbb{R}}$, $F(x, y) = f(x) + \delta_{\{u \in X : Au = y\}}(x)$ and $G(x, y) = g(y)$ for all $(x, y) \in X \times Y$. The functions F and G are proper and convex and their domains fulfill the relation

$$\mathrm{dom}\, F - \mathrm{dom}\, G = X \times (A(\mathrm{dom}\, f) - \mathrm{dom}\, g).$$

Since $\text{epi}\, F = \{(x, Ax, r) : f(x) \leq r\}$ and $\widehat{\text{epi}}(G - v(P_F^A)) = \{(x, y, r) : r \leq -G(x, y) + v(P_F^A)\} = X \times \widehat{\text{epi}}(g - v(P_F^A))$, we obtain

$$\text{epi}\, F - \widehat{\text{epi}}(G - v(P_F^A)) = X \times \left(A \times \text{id}_{\mathbb{R}}(\text{epi}\, f) - \widehat{\text{epi}}(g - v(P_F^A)) \right).$$

Moreover,

$$\inf_{(x,y) \in X \times Y} \{F(x, y) + G(x, y)\} = \inf_{x \in X} \{f(x) + (g \circ A)(x)\} = v(P_F^A).$$

On the other hand, for all $(x^*, y^*) \in X^* \times Y^*$ we have $F^*(x^*, y^*) = f^*(x^* + A^*y^*)$ and

$$G^*(x^*, y^*) = \begin{cases} g^*(y^*), & \text{if } x^* = 0, \\ +\infty, & \text{otherwise.} \end{cases}$$

Therefore

$$\sup_{\substack{x^* \in X^* \\ y^* \in Y^*}} \{-F^*(-x^*, -y^*) - G^*(x^*, y^*)\} = \sup_{y^* \in Y^*} \{-f^*(-A^*y^*) - g^*(y^*)\} = v(D_F^A).$$

We consider the following regularity conditions:

(RC_1^{FA}) | $A(\text{dom}\, f) \cap \text{qri}(\text{dom}\, g) \neq \emptyset,\ 0 \in \text{qi}(\text{dom}\, g - \text{dom}\, g)$ and
$(0, 0) \notin \text{qri}\left[\text{co}\left((A \times \text{id}_{\mathbb{R}}(\text{epi}\, f) - \widehat{\text{epi}}(g - v(P_F^A))) \cup \{(0, 0)\} \right) \right]$;

(RC_2^{FA}) | $0 \in \text{qi}\left(A(\text{dom}\, f) - \text{dom}\, g \right)$ and
$(0, 0) \notin \text{qri}\left[\text{co}\left((A \times \text{id}_{\mathbb{R}}(\text{epi}\, f) - \widehat{\text{epi}}(g - v(P_F^A))) \cup \{(0, 0)\} \right) \right]$

and

(RC_3^{FA}) | $0 \in \text{qi}\left[(A(\text{dom}\, f) - \text{dom}\, g) - (A(\text{dom}\, f) - \text{dom}\, g) \right],$
$0 \in \text{qri}\left(A(\text{dom}\, f) - \text{dom}\, g \right)$ and
$(0, 0) \notin \text{qri}\left[\text{co}\left((A \times \text{id}_{\mathbb{R}}(\text{epi}\, f) - \widehat{\text{epi}}(g - v(P_F^A))) \cup \{(0, 0)\} \right) \right].$

Similar remarks as in Lemma 2.3 can be made also for the regularity conditions (RC_i^{FA}), $i \in \{1, 2, 3\}$.

Lemma 2.4 *Under the hypotheses we work with the following statements hold:*

(i) $(RC_1^{FA}) \Rightarrow (RC_2^{FA}) \Leftrightarrow (RC_3^{FA})$;

(ii) *in case the primal problem has an optimal solution, the condition* $(0, 0) \notin \text{qri}\left[\text{co}\left((A \times \text{id}_{\mathbb{R}}(\text{epi}\, f) - \widehat{\text{epi}}(g - v(P_F^A))) \cup \{(0, 0)\} \right) \right]$ *can be equivalently written as* $(0, 0) \notin \text{qri}\left(A \times \text{id}_{\mathbb{R}}(\text{epi}\, f) - \widehat{\text{epi}}(g - v(P_F^A)) \right)$;

(iii) *if* $0 \in \text{qi}\left[(A(\text{dom}\, f) - \text{dom}\, g) - (A(\text{dom}\, f) - \text{dom}\, g) \right]$, *then the condition* $(0, 0) \notin \text{qri}\left[\text{co}\left((A \times \text{id}_{\mathbb{R}}(\text{epi}\, f) - \widehat{\text{epi}}(g - v(P_F^A))) \cup \{(0, 0)\} \right) \right]$ *is equivalent to* $(0, 0) \notin \text{qi}\left[\text{co}\left((A \times \text{id}_{\mathbb{R}}(\text{epi}\, f) - \widehat{\text{epi}}(g - v(P_F^A))) \cup \{(0, 0)\} \right) \right].$

Remark 2.13 The condition $0 \in \text{qi}\left(A(\text{dom}\, f) - \text{dom}\, g \right)$ implies relation $0 \in \text{qi}\left[(A(\text{dom}\, f) - \text{dom}\, g) - (A(\text{dom}\, f) - \text{dom}\, g) \right]$ in Lemma 2.4(iii).

Applying Theorem 2.3 for the functions F and G defined above and taking into account the above mentioned properties we obtain the following strong duality result concerning the primal-dual pair $(P_F^A) - (D_F^A)$.

Theorem 2.5 *Suppose that one of the regularity conditions (RC_i^{FA}), $i \in \{1, 2, 3\}$, is fulfilled. Then $v(P_F^A) = v(D_F^A)$ and (D_F^A) has an optimal solution.*

Remark 2.14 Let us notice that BORWEIN AND LEWIS gave in [14] some regularity conditions by means of the quasi-relative interior, in order to guarantee strong duality for the pair $(P_F^A) - (D_F^A)$. However, they considered a more restrictive case, namely that the codomain of the linear operator is finite-dimensional. We consider here a more general framework, when both of the spaces are infinite-dimensional.

2.3 Lagrange duality

Consider the optimization problem

$$(P_L) \quad \inf_{\substack{x \in S \\ g(x) \in -C}} f(x),$$

where X and Y are separated locally convex spaces, S is a non-empty convex subset of X, $f : X \to \overline{\mathbb{R}}$ is proper and convex, $C \subseteq Y$ is a non-empty convex cone, $g : X \to Y^\bullet$ is proper and C-convex and the feasible set $\mathcal{T} = \{x \in S : g(x) \in -C\}$ is assumed to be non-empty. The Lagrange dual problem associated to (P_L) is

$$(D_L) \quad \sup_{\lambda \in C^*} \inf_{x \in S} \{f(x) + \langle \lambda, g(x) \rangle\}.$$

Like in the previous section, let us recall some regularity conditions from the literature which guarantee strong duality:

$$(RC_1^L) \quad | \quad \exists x' \in \mathrm{dom}\, f \cap S \text{ such that } g(x') \in -\mathrm{int}(C);$$

(RC_2^L) | X and Y are Fréchet spaces, S is closed, f is lower semicontinuous, g is C-epi-closed and $0 \in \mathrm{int}\left(g(\mathrm{dom}\, f \cap S \cap \mathrm{dom}\, g) + C\right)$;

(RC_3^L) | X and Y are Fréchet spaces, S is closed, f is lower semicontinuous, g is C-epi-closed and $0 \in \mathrm{core}\left(g(\mathrm{dom}\, f \cap S \cap \mathrm{dom}\, g) + C\right)$;

(RC_4^L) | X and Y are Fréchet spaces, S is closed, f is lower semicontinuous, g is C-epi-closed, $0 \in \mathrm{icr}\left(g(\mathrm{dom}\, f \cap S \cap \mathrm{dom}\, g) + C\right)$ and $\mathrm{aff}\left(g(\mathrm{dom}\, f \cap S \cap \mathrm{dom}\, g) + C\right)$ is a closed linear subspace of Y

and

(RC_5^L) | X and Y are Fréchet spaces, S is closed, f is lower semicontinuous, g is C-epi-closed and $0 \in \mathrm{sqri}\left(g(\mathrm{dom}\, f \cap S \cap \mathrm{dom}\, g) + C\right)$.

Remark 2.15 The condition (RC_1^L) is the classical *Slater constraint qualification*. The above regularity conditions guarantee strong duality (cf. [21], see also [82] and [38] for stronger conditions). In case X and Y are Fréchet spaces, S is closed, f is lower semicontinuous and g is C-epi-closed, we have $(RC_1^L) \Rightarrow (RC_2^L) \Leftrightarrow (RC_3^L) \Rightarrow (RC_4^L) \Leftrightarrow (RC_5^L)$ (cf. [71], see also Remark 2.7). We refer to [82] and [38] (see also [21]) for the so-called *closedness-type regularity conditions* which ensure strong duality, too.

By using an approach due to MAGNANTI (cf. [93]), in this section we derive
from the Fenchel duality results given in the previous section duality results concerning the primal optimization problem with geometric and cone constraints and
its Lagrange dual problem.

We work in the following setting. Let X be a topological vector space and S a
non-empty subset of X. Let Y be a separated locally convex space partially ordered
by a non-empty convex cone $C \subseteq Y$. Let $f : S \to \mathbb{R}$ and $g : S \to Y$ be two functions
such that the pair $(f, g) : S \to \mathbb{R} \times Y$, defined by $(f, g)(x) = (f(x), g(x))$ for all
$x \in S$, is *convex-like* with respect to the cone $\mathbb{R}_+ \times C \subseteq \mathbb{R} \times Y$, that is the set
$(f, g)(S) + \mathbb{R}_+ \times C$ is convex. Let us notice that this property implies that the sets
$f(S) + [0, \infty)$ and $g(S) + C$ are convex (the reverse implication does not always
hold). Let us denote by $v(P_L)$ and $v(D_L)$ the optimal objective values of the primal
and the dual problem, respectively. As in the previous section, we suppose that
$v(P_L)$ is a real number.

Consider the following convex set

$$\mathcal{E}_{v(P_L)} = \{(f(x) + \alpha - v(P_L), g(x) + y) : x \in S, \alpha \geq 0, y \in C\} \subseteq \mathbb{R} \times Y.$$

Let us notice that the set $-\mathcal{E}_{v(P_L)}$ is in analogy to the *conic extension*, a notion
used by F. GIANNESSI in the theory of image space analysis (see [68]). One can
easily prove that the primal problem (P_L) has an optimal solution if and only if
$(0, 0) \in \mathcal{E}_{v(P_L)}$. Let us introduce the functions $f_1, f_2 : \mathbb{R} \times Y \to \overline{\mathbb{R}}$,

$$f_1(r, y) = \begin{cases} r, & \text{if } (r, y) \in \mathcal{E}_{v(P_L)} + (v(P_L), 0), \\ +\infty, & \text{otherwise} \end{cases}$$

and $f_2 = \delta_{\mathbb{R} \times (-C)}$, respectively. One can prove that

$$\text{dom } f_1 - \text{dom } f_2 = \mathbb{R} \times (g(S) + C). \tag{2.13}$$

Further, epi $f_1 = \{(r, y, s) \in \mathbb{R} \times Y \times \mathbb{R} : (r, y) \in \mathcal{E}_{v(P_L)} + (v(P_L), 0), r \leq s\} =$
$\{(f(x) + \alpha, g(x) + y, s) : x \in S, \alpha \geq 0, y \in C, f(x) + \alpha \leq s\}$ and $\widehat{\text{epi}}(f_2 - v(P_L)) =$
$\{(r, y, s) \in \mathbb{R} \times Y \times \mathbb{R} : s \leq -f_2(r, y) + v(P_L)\} = \{(r, y, s) \in \mathbb{R} \times Y \times \mathbb{R} : r \in$
$\mathbb{R}, y \in -C, s \leq v(P_L)\} = \mathbb{R} \times (-C) \times (-\infty, v(P_L)]$. Thus epi $f_1 - \widehat{\text{epi}}(f_2 - v(P_L)) =$
epi $f_1 + \mathbb{R} \times C \times [-v(P_L), +\infty) = \{(f(x) + \alpha + a, g(x) + y, s - v(P_L) + \varepsilon) : x \in S, \alpha \geq$
$0, a \in \mathbb{R}, y \in C, \varepsilon \geq 0, f(x) + \alpha \leq s\} = \{(f(x) + \alpha + a, g(x) + y, f(x) + \alpha + \varepsilon - v(P_L)) :$
$x \in S, \alpha \geq 0, a \in \mathbb{R}, y \in C, \varepsilon \geq 0\}$ and this means that

epi $f_1 - \widehat{\text{epi}}(f_2 - v(P_L)) = \mathbb{R} \times \{(g(x) + y, f(x) + \alpha - v(P_L)) : x \in S, \alpha \geq 0, y \in C\}.$

Moreover, as pointed out by MAGNANTI (cf. [93]), we have

$$\inf_{(r,y) \in \mathbb{R} \times Y} \{f_1(r, y) + f_2(r, y)\} = \inf_{\substack{x \in S \\ g(x) \in -C}} f(x) = v(P_L) \tag{2.14}$$

and

$$\sup_{(r^*, y^*) \in \mathbb{R} \times Y^*} \{-f_1^*(-r^*, -y^*) - f_2^*(r^*, y^*)\} = \sup_{\lambda \in C^*} \inf_{x \in S} \{f(x) + \langle \lambda, g(x) \rangle\} = v(D_L). \tag{2.15}$$

By considering the relations (RC_i^F), $i \in \{6, 7, 8\}$, for the pair of functions (f_1, f_2)
and using the approach due to MAGNANTI, one can derive the following regularity
conditions by means of the quasi interior and quasi-relative interior:

$$(RC_6^L) \quad \Bigg| \quad \begin{array}{l} \text{cl}(C - C) = Y, \ \exists x' \in S \text{ such that } g(x') \in -\text{qri}(C) \text{ and} \\ (0, 0) \notin \text{qri}\left[\text{co}(\mathcal{E}_{v(P_L)} \cup \{(0, 0)\})\right]; \end{array}$$

$$(RC_7^L) \quad | \quad 0 \in \text{qi}(g(S) + C) \text{ and } (0,0) \notin \text{qri}\left[\text{co}(\mathcal{E}_{v(P_L)} \cup \{(0,0)\})\right]$$

and

$$(RC_8^L) \quad \left| \begin{array}{l} 0 \in \text{qi}\left[(g(S) + C) - (g(S) + C)\right], \ 0 \in \text{qri}(g(S) + C) \text{ and} \\ (0,0) \notin \text{qri}\left[\text{co}(\mathcal{E}_{v(P_L)} \cup \{(0,0)\})\right]. \end{array} \right.$$

Remark 2.16 The fact that the condition (RC_6^F) written for the pair (f_1, f_2) is equivalent to (RC_6^L) follows from the relation $\text{qri}(C) + C = \text{qri}(C)$, which is a direct consequence of Proposition 2.3(v). Let us notice that the condition (RC_6^L) is considered also in [103] in case Y is a normed space.

We study in the following the relations between the above considered regularity conditions.

Lemma 2.5 *Suppose that* $\text{cl}(C - C) = Y$ *and* $\exists x' \in S$ *such that* $g(x') \in - \text{qri}(C)$. *Then the following assertions are true*

(i) $0 \in \text{qi}(g(S) + C)$;

(ii) $\text{cl}\left[\text{cone}\left(\text{qri}(g(S) + C)\right)\right] = Y.$

Proof. (i) The condition $\text{cl}(C - C) = Y$ implies $0 \in \text{qi}(C - C)$, while the Slater-type condition $g(x') \in - \text{qri}\, C$ ensures that $\text{qri}(C) \cap (-g(S) - C) \neq \emptyset$. Hence, by Lemma 2.1(i) we obtain $0 \in \text{qi}\left(C - (-g(S) - C)\right)$ which is nothing else than $0 \in \text{qi}(g(S) + C)$.

(ii) From (i) it follows that $0 \in \text{qri}(g(S) + C)$, hence $\text{qri}(g(S) + C) \neq \emptyset$. Applying Proposition 2.3 (ix) we get $\text{cl}\left[\text{cone}(\text{qri}(g(S) + C))\right] = \text{cl}\left(\text{cone}(g(S) + C)\right) = Y$, the later equality being a consequence of (i). $\qquad \square$

Similar remarks as in Lemma 2.3 can be made also for the regularity conditions (RC_i^L), $i \in \{6, 7, 8\}$.

Lemma 2.6 *Under the hypotheses we work with the following statements hold:*

(i) $(RC_6^L) \Rightarrow (RC_7^L) \Leftrightarrow (RC_8^L)$;

(ii) in case the primal problem has an optimal solution, the condition $(0,0) \notin \text{qri}\left[\text{co}(\mathcal{E}_{v(P_L)} \cup \{(0,0)\})\right]$ *can be equivalently written as* $(0,0) \notin \text{qri}\left(\mathcal{E}_{v(P_L)}\right)$;

(iii) if $0 \in \text{qi}\left[(g(S) + C) - (g(S) + C)\right]$, *then the condition* $(0,0) \notin \text{qri}\left[\text{co}(\mathcal{E}_{v(P_L)} \cup \{(0,0)\})\right]$ *is equivalent to* $(0,0) \notin \text{qi}\left[\text{co}(\mathcal{E}_{v(P_L)} \cup \{(0,0)\})\right]$.

Remark 2.17 Let us notice that a sufficient condition for the fulfillment of $0 \in \text{qi}\left[(g(S) + C) - (g(S) + C)\right]$ in Lemma 2.6(iii) is $0 \in \text{qi}(g(S) + C)$. This is a direct consequence of the inclusion $g(S) + C \subseteq (g(S) + C) - (g(S) + C)$.

Remark 2.18 Similarly as in Remark 2.10(c), one can prove the implication

$$(0,0) \in \text{qi}\left(\text{co}\left(\mathcal{E}_{v(P_L)} \cup \{(0,0)\}\right)\right) \Rightarrow 0 \in \text{qi}(g(S) + C).$$

Let us mention that one cannot substitute in (RC_i^L), $i \in \{6, 7, 8\}$, the condition $(0,0) \notin \text{qri}\left[\text{co}\left(\mathcal{E}_{v(P_L)} \cup \{(0,0)\}\right)\right]$ by $0 \notin \text{qi}(g(S) + C)$, since this would be in contradiction with the other assumptions (see Lemma 2.5 and Lemma 2.6).

Applying Theorem 2.3 for the functions f_1 and f_2 defined above and taking into account the above presented approach of MAGNANTI we obtain the following strong duality result concerning the primal-dual pair $(P_L) - (D_L)$.

Theorem 2.6 *Suppose that one of the regularity conditions* (RC_i^L), $i \in \{6,7,8\}$, *is fulfilled. Then* $v(P_L) = v(D_L)$ *and* (D_L) *has an optimal solution.*

Let us give in the following an example which underlines the applicability of the above strong duality theorem (for another example we refer to [26, Example 4.1]).

Example 2.8 Consider again the Hilbert space $\ell^2(\mathbb{N})$ and the following setting: $S = C = \ell_+^2$, $f : \ell_+^2 \to \mathbb{R}$, $f(x) = \langle c, x \rangle$ and $g : \ell_+^2 \to \ell^2$, $g(x) = x - x^0$, where $c, x^0 \in \ell_+^2$ are arbitrary chosen such that $x_n^0 > 0$ for all $n \in \mathbb{N}$. The feasible set of the primal problem is $\mathcal{T} = \ell_+^2 \cap (x^0 - \ell_+^2) \neq \emptyset$ and it holds

$$v(P_L) = \inf_{x \in \mathcal{T}} \langle c, x \rangle = 0,$$

while $\overline{x} = 0$ is an optimal solution of the primal problem. The condition $\mathrm{cl}(C - C) = \ell^2$ is obviously satisfied and we have that (cf. Example 2.1) $\{x \in S : g(x) \in -\mathrm{qri}(C)\} = \{x = (x_n)_{n \in \mathbb{N}} \in \ell^2 : 0 \leq x_n < x_n^0 \ \forall n \in \mathbb{N}\}$. This is a non-empty set, hence the Slater-type condition is also fulfilled. We prove in the following that $(0,0) \notin \mathrm{qi}\left(\mathcal{E}_{v(P_L)}\right)$. An arbitrary element $(r^*, x^*) \in \mathbb{R} \times \ell^2$ belongs to $N_{\mathcal{E}_{v(P_L)}}(0,0)$ if and only if

$$r^*(\langle c, x \rangle + \alpha) + \langle x^*, x - x^0 + y \rangle \leq 0 \ \forall x \in \ell_+^2 \ \forall \alpha \geq 0 \ \forall y \in \ell_+^2.$$

One can observe that $(-1,0) \in N_{\mathcal{E}_{v(P_L)}}(0,0)$, which ensures that $N_{\mathcal{E}_{v(P_L)}}(0,0) \neq \{(0,0)\}$. By using Proposition 2.2 we obtain the conclusion. By Lemma 2.6, the condition (RC_6^L) is fulfilled and hence strong duality holds. Let us notice that, since $g(S) + C = \ell_+^2 - x^0$, none of the regularity conditions (RC_i^L), $i \in \{1,2,3,4,5\}$, presented in this section can be applied to this problem (see Example 2.1). The optimal objective value of the Lagrange dual problem is

$$v(D_L) = \sup_{\lambda \in \ell_+^2} \inf_{x \in \ell_+^2} \{\langle c, x \rangle + \langle \lambda, x - x^0 \rangle\}$$

$$= \sup_{\lambda \in \ell_+^2} \left\{ -\langle \lambda, x^0 \rangle + \inf_{x \in \ell_+^2} \langle c + \lambda, x \rangle \right\} = \sup_{\lambda \in \ell_+^2} \langle -\lambda, x^0 \rangle = 0$$

and $\overline{\lambda} = 0$ is an optimal solution of the dual.

The following example considered by DANIELE AND GIUFFRÈ in [60] shows that if the condition $(0,0) \notin \mathrm{qri}\left[\mathrm{co}(\mathcal{E}_{v(P_L)} \cup \{(0,0)\})\right]$ is removed, the strong duality result may fail.

Example 2.9 Let be $X = S = Y = \ell^2$ and $C = \ell_+^2$. Take $f : \ell^2 \to \mathbb{R}$, $f(x) = \langle c, x \rangle$, where $c = (c_n)_{n \in \mathbb{N}}$, $c_n = (1/n)$ for all $n \in \mathbb{N}$ and $g : \ell^2 \to \ell^2$, $g(x) = -Ax$, where $(Ax)_n = (1/2^n)x_n$ for all $n \in \mathbb{N}$. Then $\mathcal{T} = \{x \in \ell^2 : Ax \in \ell_+^2\} = \ell_+^2$. It holds $\mathrm{cl}(\ell_+^2 - \ell_+^2) = \ell^2$ and $\mathrm{qri}(\ell_+^2) = \{x = (x_n)_{n \in \mathbb{N}} \in \ell^2 : x_n > 0 \ \forall n \in \mathbb{N}\} \neq \emptyset$ and one can easily find an $\overline{x} \in \ell^2$ with $g(\overline{x}) \in -\mathrm{qri}(\ell_+^2)$. We also have that

$$v(P_L) = \inf_{x \in \mathcal{T}} \langle c, x \rangle = 0$$

and $x = 0$ is an optimal solution of the primal problem. On the other hand, for $\lambda \in C^* = \ell_+^2$, it holds

$$\inf_{x \in S} \{f(x) + \langle \lambda, g(x) \rangle\} = \inf_{x \in \ell^2} \{\langle c, x \rangle + \langle \lambda, g(x) \rangle\}$$

$$= \inf_{x = (x_n)_{n \in \mathbb{N}} \in \ell^2} \left(\sum_{n=1}^{\infty} \frac{1}{n} x_n - \sum_{n=1}^{\infty} \lambda_n \frac{1}{2^n} x_n \right) = \inf_{(x_n)_{n \in \mathbb{N}} \in \ell^2} \sum_{n=1}^{\infty} \left(\frac{1}{n} - \frac{\lambda_n}{2^n} \right) x_n$$

$$= \begin{cases} 0, & \text{if } \lambda_n = \frac{2^n}{n} \ \forall n \in \mathbb{N}, \\ -\infty, & \text{otherwise.} \end{cases}$$

Since $(2^n/n)_{n\in\mathbb{N}}$ does not belong to ℓ^2, we obtain $v(D_L) = -\infty$, hence strong duality fails.

Moreover, it is not surprising that strong duality does not hold, since not all the conditions in (RC_i^L), $i \in \{6,7,8\}$, are fulfilled. This follows as one can prove that $(0,0) \in \text{qi}\,(\mathcal{E}_{v(P_L)})$. Indeed, take an arbitrary element $(r^*, x^*) \in N_{\mathcal{E}_{v(P_L)}}(0,0)$ with $x^* = (x_n^*)_{n\in\mathbb{N}} \in \ell^2$ and $r^* \in \mathbb{R}$. Then we have

$$r^*(\langle c, x \rangle + \alpha) + \langle x^*, g(x) + y \rangle \le 0 \ \forall x \in \ell^2 \ \forall \alpha \ge 0 \ \forall y \in \ell^2_+, \tag{2.16}$$

that is

$$r^*\left(\sum_{n=1}^{\infty} \frac{1}{n} x_n + \alpha\right) + \sum_{n=1}^{\infty} x_n^*\left(-\frac{1}{2^n} x_n + y_n\right) \le 0$$

$$\forall x = (x_n)_{n\in\mathbb{N}} \in \ell^2 \ \forall \alpha \ge 0 \ \forall y = (y_n)_{n\in\mathbb{N}} \in \ell^2_+.$$

Taking $\alpha = 0$ and $y_n = 0$ for all $n \in \mathbb{N}$ in the relation above we get

$$\sum_{n=1}^{\infty} \left(r^* \frac{1}{n} - \frac{1}{2^n} x_n^*\right) x_n \le 0 \ \forall x = (x_n)_{n\in\mathbb{N}} \in \ell^2,$$

which implies $x_n^* = r^*(2^n/n)$ for all $n \in \mathbb{N}$. Since $x^* \in \ell^2$, we must have $r^* = 0$ and hence $x^* = 0$. Thus $N_{\mathcal{E}_{v(P_L)}}(0,0) = \{(0,0)\}$ and so $(0,0) \in \text{qi}\,(\mathcal{E}_{v(P_L)})$ (cf. Proposition 2.2).

As in the previous section, we show that in general the condition (RC_7^L) (and automatically also (RC_8^L), see Lemma 2.6(i)) is weaker than (RC_6^L).

Example 2.10 Consider the following setting: $X = Y = \ell^2(\mathbb{R})$, $S = C = \ell^2_+(\mathbb{R})$ and the functions $f : \ell^2_+(\mathbb{R}) \to \mathbb{R}$, $g : \ell^2_+(\mathbb{R}) \to \ell^2(\mathbb{R})$ defined by $f(s) = \|s\|$ and $g(s) = -s$, for all $s \in \ell^2_+(\mathbb{R})$, respectively. For the primal problem we have

$$v(P_L) = \inf_{x \in \ell^2_+(\mathbb{R})} \|s\| = 0$$

and $s = 0$ is an optimal solution. Since $\text{qri}\,(\ell^2_+(\mathbb{R})) = \emptyset$ (cf. Example 2.2), the condition (RC_6^L) fails. Further, $g(S) + C = -\ell^2_+(\mathbb{R}) + \ell^2_+(\mathbb{R}) = \ell^2(\mathbb{R})$, hence $0 \in \text{qi}(g(S) + C)$. Like in Example 2.8 one can prove that $(0,0) \notin \text{qi}\,(\mathcal{E}_{v(P_L)})$, thus the condition (RC_7^L) is fulfilled (see also Lemma 2.6), hence strong duality holds (cf. Theorem 2.6). The optimal objective value of the dual problem is

$$v(D_L) = \sup_{\lambda \in \ell^2_+(\mathbb{R})} \inf_{s \in \ell^2_+(\mathbb{R})} \{\|s\| - \langle \lambda, s \rangle\}.$$

For every $\lambda \in \ell^2_+(\mathbb{R})$ the inner infimum in the above relation can be written as (cf. [149, Theorem 2.8.7])

$$\inf_{s \in \ell^2_+(\mathbb{R})} \{\|s\| - \langle \lambda, s \rangle\} = - \sup_{s \in \ell^2_+(\mathbb{R})} \{\langle \lambda, s \rangle - \|s\|\} = -(\|\cdot\| + \delta_{\ell^2_+(\mathbb{R})})^*(\lambda)$$

$$= -(\delta_{\overline{B}(0,1)} \square \delta_{-\ell^2_+(\mathbb{R})})(\lambda) = -\delta_{\overline{B}(0,1) - \ell^2_+(\mathbb{R})}(\lambda),$$

where $\overline{B}(0,1)$ is the closed unit ball of $(\ell^2(\mathbb{R}))^* = \ell^2(\mathbb{R})$. We get $v(D_L) = 0$ and every $\lambda \in \ell^2_+(\mathbb{R}) \cap (\overline{B}(0,1) - \ell^2_+(\mathbb{R}))$ is an optimal solution of the dual (in particular also $\overline{\lambda} = 0$).

Remark 2.19 Let us mention that JEYAKUMAR AND WOLKOWICZ have introduced in [85] some regularity conditions in terms of the quasi-relative interior, in order to guarantee Lagrange duality. However, most of these conditions require the interior of the ordering cone to be non-empty, which is not always fulfilled, as we pointed out at the beginning of the chapter.

For the rest of this section we revisit some results recently given on this topic which, unfortunately, have either superfluous or contradictory hypotheses. One can overcome these drawbacks by using the results presented in this chapter.

A regularity condition for strong duality for the pair $(P_L) - (D_L)$ was proposed by CAMMAROTO AND DI BELLA in [55, Theorem 2.2]:

Let X be a topological vector space and let S be a non-empty subset of X; let $(Y, \|\cdot\|)$ be a normed space partially ordered by a convex cone C; let $f : S \to \mathbb{R}$ and $g : S \to Y$ be two functions such that the function $(f, g) : S \to \mathbb{R} \times Y$ defined above is convex-like with respect to the cone $\mathbb{R}_+ \times C$ of $\mathbb{R} \times Y$, $\mathrm{qri}(g(S) + C) \neq \emptyset$ and $\mathrm{cl}\left[\mathrm{cone}(\mathrm{qri}(g(S) + C))\right]$ is not a linear subspace of Y. Let the set $T = \{x \in S : g(x) \in -C\}$ be non-empty. In addition, suppose that $\mathrm{qri}(C) \neq \emptyset$ and $\mathrm{cl}(C-C) = Y$. If the problem (P_L) is solvable and there exists $x' \in S$ with $g(x') \in -\mathrm{qri}(C)$, then the problem (D_L) is also solvable and the extrema of the problems are equal.

Lemma 2.5 shows that this theorem finds no application, since the hypotheses are contradictory. Let us notice that [55, Theorem 2.2] was used also in [69, 70] for generalized complementarity problems.

In [61] DANIELE, GIUFFRÈ, IDONE AND MAUGERI considered the following notion regarding the problem (P_L): we say that *Assumption S* is fulfilled at $x_0 \in T$ if

$$(\textit{Assumption S}) \qquad T_{\widetilde{M}}(f(x_0), 0) \cap ((-\infty, 0) \times 0) = \emptyset,$$

where

$$\widetilde{M} := \{(f(x) + \alpha, g(x) + y) : x \in S \setminus T, \alpha \geq 0, y \in C\}.$$

They also formulated the following strong duality theorem (cf. [61, Theorem 4]):

Let X be a topological vector space and S a non-empty subset of X; let $(Y, \|\cdot\|)$ be a normed space partially ordered by a convex cone C. Let $f : S \to \mathbb{R}$ and $g : S \to Y$ be two functions such that the function $(f, g) : S \to \mathbb{R} \times Y$ defined above is convex-like with respect to the cone $\mathbb{R}_+ \times C$ of $\mathbb{R} \times Y$. Let the set $T = \{x \in S : g(x) \in -C\}$ be non-empty and let us assume that $\mathrm{qri}\, C \neq \emptyset$, $\mathrm{cl}(C - C) = Y$ and there exists $\overline{x} \in S$ with $g(\overline{x}) \in -\mathrm{qri}\, C$. Then if the problem (P_L) is solvable and Assumption S is fulfilled at the extremal solution $x_0 \in T$ to the problem (P_L), also the problem (D_L) is solvable, the extrema values of both problems are equal and it results

$$\langle \overline{u}, g(x_0) \rangle = 0,$$

where $\overline{u} \in C^*$ is the extremal point of the problem (D_L).

Remark 2.20 We emphasize that *Assumption S* is fulfilled at an optimal solution of the primal problem if and only for the pair $(P_L) - (D_L)$ strong duality holds (see [26, Corollary 3.1]). This means that the other assumptions which involve also the quasi-relative interior in the result presented above are superfluous. Moreover, for the primal optimization problem with both cone and equality constraints a similar *Assumption S* is used in [60, Theorem 3.1]. Again, this condition is equivalent to strong duality, making the other conditions considered by the authors superfluous (cf. [26, Corollary 3.1]). For a detailed proof of the above considerations concerning *Assumption S* we refer to [26]. Finally, let us mention that the *Assumption S,* together with other conditions expressed via the quasi-relative interior, are used in other papers too, like [59, 101].

Remark 2.21 A valuable strong duality theorem for the primal optimization problem with cone and equality constraints and its Lagrange dual problem is given in [72] by means of the quasi interior and quasi-relative interior.

Remark 2.22 A comparison of the regularity conditions introduced in this chapter by means of the quasi interior and quasi-relative interior with the classical ones mentioned in the last two sections of this chapter is provided in [25].

Chapter 3

Sequential optimality conditions in convex optimization

The theory developed in this chapter is motivated by the followings. Consider the convex optimization problem

$$(P_F^0) \quad \inf_{x \in D} f(x),$$

where $f : X \to \overline{\mathbb{R}}$ is a proper and convex function, X is a separated locally convex space and D is a non-empty convex subset of X. The celebrated Pshenichnyi-Rockafellar Lemma (see [118, 122, 149]) provides a necessary and sufficient optimality condition for the problem (P_F^0), whenever a regularity condition is fulfilled: in case $\operatorname{dom} f \cap \operatorname{int}(D) \neq \emptyset$ (or f is continuous at some $x_0 \in \operatorname{dom} f \cap D$), an element $a \in \operatorname{dom} f \cap D$ is an optimal solution of the problem (P_F^0) if and only if $0 \in \partial f(a) + N_D(a)$. This is a very important result in convex optimization with many applications. Nevertheless, it has some disadvantages. First of all, a can be a minimizer of f on D even if $0 \notin \partial f(a) + N_D(a)$ (as, for instance, the set $\partial f(a)$ could be empty; see [86] for such an example). Moreover, the regularity conditions are not always fulfilled even in the finite-dimensional case. The same disadvantages arise also in the case of optimization problems with geometric and cone constraints.

Trying to eliminate these drawbacks, many mathematicians have given optimality conditions that do not require any regularity condition. With respect to the problem (P_F^0), a nice generalization of the Pshenichnyi-Rockafellar Lemma was recently given by JEYAKUMAR AND WU in [86]. It is stated in terms of a sequence of ε-subdifferentials and ε-normal sets and provides a necessary and sufficient optimality condition without asking the fulfillment of any regularity condition.

For the problem with geometric and cone constraints various modified Lagrange multiplier conditions without regularity conditions have been given in the literature (cf. [8, 9, 18, 19, 58, 81, 89]). Moreover, in [83] and [88], several qualification free sequential optimality conditions for this problem are introduced . In [138] THIBAULT gave a sequential form of the Lagrange multiplier rule in the case the cone which appears in the constraint set is convex, closed and normal. Other sequential characterizations can be found in literature in [79, 106, 137, 139].

Motivated by these considerations we give in Section 3.1 sequential optimality conditions without any regularity condition for the general convex optimization problem

$$(P_\Phi) \quad \inf_{x \in X} \Phi(x, 0),$$

where $\Phi : X \times Y \to \overline{\mathbb{R}}$, the so-called *perturbation function*, is proper, convex and lower semicontinuous and X, Y are Banach spaces, X being supposed reflexive. This sequential characterization is obtained by using the properties of the infimal value function of the conjugate Φ^* and the formula for the epigraph of a conjugate function written in terms of the ε-subdifferential (see (1. 3)). Combining the above condition with a version of the Brøndsted-Rockafellar Theorem (see Theorem 1.1), we obtain another qualification free sequential characterization of optimal solutions involving the classical (convex) subdifferential.

In the last three sections of this chapter we consider particular instances of the general results, rediscovering and, in many situations, even improving some sequential optimality conditions given in the literature by JEYAKUMAR AND WU and THIBAULT, respectively. The main results of this chapter are Theorem 3.1 and Theorem 3.2 and the theory presented here is based on [28, 29].

3.1   A general approach via perturbation theory

Consider $(X, \|\cdot\|)$ a reflexive Banach space, $(Y, \|\cdot\|)$ a Banach space and $(X^*, \|\cdot\|_*)$, $(Y^*, \|\cdot\|_*)$ their topological dual spaces, respectively. Although the spaces X, Y and X^*, Y^*, respectively, are endowed with different norms, we use the same notations for these as there is no danger of confusion. Let $\{x_n^* : n \in \mathbb{N}\}$ be a sequence in X^*. We write $x_n^* \xrightarrow{\omega^*} 0$ $(x_n^* \xrightarrow{\|\cdot\|_*} 0)$ for the case when x_n^* converges to 0 in the weak* (strong) topology on X^*. We make the following convention: if in a certain property for the convergence in the dual space we write $x_n^* \to 0$ $(n \to +\infty)$, we understand that the property holds no matter which of the two topologies (weak* or strong) is used. The following property will be frequently used in this chapter:

$$\text{if } x_n^* \to 0 \text{ and } x_n \to a \ (n \to +\infty), \text{ then } \langle x_n^*, x_n \rangle \to 0 \ (n \to +\infty),$$

where $\{x_n : n \in \mathbb{N}\} \subseteq X$, $a \in X$ and $x_n \to a$ $(n \to +\infty)$ means $\|x_n - a\| \to 0$ $(n \to +\infty)$, that is the convergence in the topology induced by the norm on X. On $X \times Y$ we use the norm $\|(x, y)\| = \sqrt{\|x\|^2 + \|y\|^2}$, for $(x, y) \in X \times Y$. The norm on $X^* \times Y^*$ is defined analogously.

Let $\Phi : X \times Y \to \overline{\mathbb{R}}$ be a given function. The so-called perturbation function Φ plays a determinant role in the duality theory as it can be used for constructing a dual problem to a given primal optimization problem. More precisely, the dual problem is defined by using the conjugate of Φ (we refer to [21, 36, 62, 149] for a comprehensive study of the perturbation theory). The classical duality approaches, like Fenchel duality and Lagrange duality, can be seen as particular cases of this general theory.

In this section we give sequential optimality conditions for the general optimization problem

$$(P_\Phi) \quad \inf_{x \in X} \Phi(x, 0).$$

To this end we consider the *infimal value* function $\eta : X^* \to \overline{\mathbb{R}}$ of the conjugate Φ^* defined by $\eta(x^*) = \inf_{y^* \in Y^*} \Phi^*(x^*, y^*)$ for all $x^* \in X^*$. Let us notice that, since Φ^* is a convex function on $X^* \times Y^*$, η is a convex function on X^*. We begin our investigation by establishing a result which holds also in the framework of separated locally convex spaces.

Lemma 3.1 *Let* $\Phi : X \times Y \to \overline{\mathbb{R}}$ *be a proper, convex and lower semicontinuous function such that* $0 \in \mathrm{pr}_Y(\mathrm{dom}\,\Phi)$. *Then* $a \in \mathrm{dom}\,\Phi(\cdot, 0)$ *is an optimal solution of the problem* (P_Φ) *if and only if* $(0, -\eta^*(a)) \in \mathrm{cl}_{w^* \times \mathcal{R}}(\mathrm{epi}\,\eta)$.

Proof. One can see that $\operatorname{dom}\eta \neq \emptyset$ and $\eta^*(x) = (\Phi^*)^*(x,0) = \Phi(x,0)$ for all $x \in X$. We get that η^* is proper, hence $\operatorname{cl}\eta$ is also proper and $\eta^{**} = \operatorname{cl}\eta$. Then $a \in \operatorname{dom}\Phi(\cdot,0)$ is an optimal solution of (P_Φ) if and only if a is an optimal solution of the problem

$$(P_\eta) \quad \inf_{x \in X} \eta^*(x).$$

The later is equivalent to $0 \in \partial\eta^*(a)$, which is nothing else than $\eta^{**}(0) + \eta^*(a) \leq 0$. This is the same with $(\operatorname{cl}_{w^*}\eta)(0) = \eta^{**}(0) \leq -\eta^*(a)$, which can be reformulated as $(0, -\eta^*(a)) \in \operatorname{epi}(\operatorname{cl}_{w^*}\eta) = \operatorname{cl}_{w^* \times \mathcal{R}}(\operatorname{epi}\eta)$. $\qquad\square$

By using Lemma 3.1 and formula (1. 3) one can give now general sequential optimality conditions for the problem (P_Φ) involving ε-subdifferentials.

Theorem 3.1 *Let $\Phi : X \times Y \to \overline{\mathbb{R}}$ be a proper, convex and lower semicontinuous function such that $0 \in \operatorname{pr}_Y(\operatorname{dom}\Phi)$. The following statements are equivalent:*

(i) $a \in \operatorname{dom}\Phi(\cdot,0)$ is an optimal solution of the problem (P_Φ);

(ii) there exist sequences $\{\varepsilon_n\} \downarrow 0$ and $(x_n^*, y_n^*) \in \partial_{\varepsilon_n}\Phi(a,0)$ such that $x_n^* \xrightarrow{\|\cdot\|_*} 0$ $(n \to +\infty)$;

(iii) there exist sequences $\{\varepsilon_n\} \downarrow 0$ and $(x_n^*, y_n^*) \in \partial_{\varepsilon_n}\Phi(a,0)$ such that $x_n^* \xrightarrow{w^*} 0$ $(n \to +\infty)$.

Proof. $(i) \Rightarrow (ii)$ Suppose that $a \in \operatorname{dom}\Phi(\cdot,0)$ is an optimal solution of the problem (P_Φ). Applying the previous lemma, we have $(0, -\eta^*(a)) \in \operatorname{cl}_{w^* \times \mathcal{R}}(\operatorname{epi}\eta)$. Since η is a convex function and X is a reflexive Banach space we have

$$\operatorname{cl}_{w^* \times \mathcal{R}}(\operatorname{epi}\eta) = \operatorname{cl}_{\|\cdot\|_* \times \mathcal{R}}(\operatorname{epi}\eta).$$

Hence $\exists (x_n^*, r_n) \in X^* \times \mathbb{R}$ such that $\eta(x_n^*) \leq r_n, x_n^* \xrightarrow{\|\cdot\|_*} 0$ and $r_n \to -\eta^*(a)$ $(n \to +\infty)$. The inequality $\eta(x_n^*) \leq r_n$ yields $\inf_{y^* \in Y^*}\Phi^*(x_n^*, y^*) < r_n + 1/n$ for all $n \in \mathbb{N}$, so there exists a sequence $\{y_n^* : n \in \mathbb{N}\} \subseteq Y^*$ such that $\Phi^*(x_n^*, y_n^*) < r_n + 1/n$ for all $n \in \mathbb{N}$, thus $(x_n^*, y_n^*, r_n + 1/n) \in \operatorname{epi}\Phi^*$ for all $n \in \mathbb{N}$. As $(a, 0) \in \operatorname{dom}\Phi$, we get by (1. 3)

$$\operatorname{epi}\Phi^* = \bigcup_{\varepsilon \geq 0}\left\{(x^*, y^*, \langle x^*, a\rangle + \varepsilon - \Phi(a,0)) : (x^*, y^*) \in \partial_\varepsilon\Phi(a,0)\right\}.$$

Since $(x_n^*, y_n^*, r_n + 1/n) \in \operatorname{epi}\Phi^*$ for all $n \in \mathbb{N}$, there exists a sequence $\{\varepsilon_n : n \in \mathbb{N}\} \subseteq \mathbb{R}_+$ such that $r_n + 1/n = \langle x_n^*, a\rangle + \varepsilon_n - \Phi(a,0)$, $(x_n^*, y_n^*) \in \partial_{\varepsilon_n}\Phi(a,0)$, $x_n^* \xrightarrow{\|\cdot\|_*} 0$ $(n \to +\infty)$. As $r_n \to -\eta^*(a) = -\Phi(a,0)$ $(n \to +\infty)$, from the last equality we conclude that $\varepsilon_n \to 0$ $(n \to +\infty)$.

The implication $(ii) \Rightarrow (iii)$ is trivial.

$(iii) \Rightarrow (i)$ If there exist sequences $\{\varepsilon_n\} \downarrow 0$ and $(x_n^*, y_n^*) \in \partial_{\varepsilon_n}\Phi(a,0)$ such that $x_n^* \xrightarrow{w^*} 0$ $(n \to +\infty)$, then using the definition of the ε-subdifferential of a function we get

$$\Phi(x,y) - \Phi(a,0) \geq \langle x_n^*, x - a\rangle + \langle y_n^*, y\rangle - \varepsilon_n \ \forall (x,y) \in X \times Y \ \forall n \in \mathbb{N}.$$

We obtain

$$\Phi(x,0) - \Phi(a,0) \geq \langle x_n^*, x - a\rangle - \varepsilon_n \ \forall x \in X \ \forall n \in \mathbb{N}.$$

Passing to the limit as $n \to +\infty$, we get $\Phi(x,0) - \Phi(a,0) \geq 0$ for all $x \in X$, hence $a \in \operatorname{dom}\Phi(\cdot,0)$ is an optimal solution of the problem (P_Φ). $\qquad\square$

Combining this result with the Brøndsted-Rockafellar Theorem (Theorem 1.1) we get necessary and sufficient sequential optimality conditions by means of the (convex) subdifferential.

Theorem 3.2 Let $\Phi : X \times Y \to \overline{\mathbb{R}}$ be a proper, convex and lower semicontinuous function such that $0 \in \mathrm{pr}_Y(\mathrm{dom}\,\Phi)$. The following statements are equivalent:

(i) $a \in \mathrm{dom}\,\Phi(\cdot, 0)$ is an optimal solution of the problem (P_Φ);

(ii) there exist sequences $(x_n, y_n) \in \mathrm{dom}\,\Phi$ and $(x_n^*, y_n^*) \in \partial\Phi(x_n, y_n)$ such that

$$x_n^* \xrightarrow{\|\cdot\|_*} 0, \; x_n \to a, \; y_n \to 0 \; (n \to +\infty) \; \text{and}$$

$$\Phi(x_n, y_n) - \langle y_n^*, y_n \rangle - \Phi(a, 0) \to 0 \; (n \to +\infty);$$

(iii) there exist sequences $(x_n, y_n) \in \mathrm{dom}\,\Phi$ and $(x_n^*, y_n^*) \in \partial\Phi(x_n, y_n)$ such that

$$x_n^* \xrightarrow{w^*} 0, \; x_n \to a, \; y_n \to 0 \; (n \to +\infty) \; \text{and}$$

$$\Phi(x_n, y_n) - \langle y_n^*, y_n \rangle - \Phi(a, 0) \to 0 \; (n \to +\infty).$$

Proof. As $(ii) \Rightarrow (iii)$ is always true, we prove only the implications $(i) \Rightarrow (ii)$ and $(iii) \Rightarrow (i)$.

$(i) \Rightarrow (ii)$ Suppose that $a \in \mathrm{dom}\,\Phi(\cdot, 0)$ is an optimal solution of the problem (P_Φ). By Theorem 3.1 there exist $\{\varepsilon_n\} \downarrow 0$ and $(\overline{x_n^*}, \overline{y_n^*}) \in \partial_{\varepsilon_n}\Phi(a, 0)$ such that $\overline{x_n^*} \xrightarrow{\|\cdot\|_*} 0$ $(n \to +\infty)$. Applying Theorem 1.1 we get that for all $n \in \mathbb{N}$ there exist $(x_n, y_n) \in X \times Y$ and $(x_n^*, y_n^*) \in \partial\Phi(x_n, y_n)$ such that

$$\|(x_n, y_n) - (a, 0)\| \leq \sqrt{\varepsilon_n}, \|(x_n^*, y_n^*) - (\overline{x_n^*}, \overline{y_n^*})\|_* \leq \sqrt{\varepsilon_n}$$

and

$$|\Phi(x_n, y_n) - \langle (x_n^*, y_n^*), (x_n, y_n) - (a, 0) \rangle - \Phi(a, 0)| \leq 2\varepsilon_n,$$

from which we obtain $x_n^* \xrightarrow{\|\cdot\|_*} 0$, $x_n \to a$, $y_n \to 0$ $(n \to +\infty)$ and $\Phi(x_n, y_n) - \langle x_n^*, x_n - a \rangle - \langle y_n^*, y_n \rangle - \Phi(a, 0) \to 0$ $(n \to +\infty)$. Since $\langle x_n^*, x_n - a \rangle \to 0$ $(n \to +\infty)$, the desired result follows.

$(iii) \Rightarrow (i)$ Assume that there exist sequences $(x_n, y_n) \in \mathrm{dom}\,\Phi$, $(x_n^*, y_n^*) \in \partial\Phi(x_n, y_n)$ such that $x_n^* \xrightarrow{w^*} 0$, $x_n \to a$, $y_n \to 0$ $(n \to +\infty)$ and $\Phi(x_n, y_n) - \langle y_n^*, y_n \rangle - \Phi(a, 0) \to 0$ $(n \to +\infty)$. Since $(x_n^*, y_n^*) \in \partial\Phi(x_n, y_n)$, we have $\Phi(x, y) \geq \Phi(x_n, y_n) + \langle (x_n^*, y_n^*), (x - x_n, y - y_n) \rangle$ for all $(x, y) \in X \times Y$ and all $n \in \mathbb{N}$. Consequently, for every $x \in X$ the following inequality is true

$$\Phi(x, 0) - \Phi(a, 0) \geq \Phi(x_n, y_n) - \langle y_n^*, y_n \rangle - \Phi(a, 0) + \langle x_n^*, x - x_n \rangle \; \forall n \in \mathbb{N}.$$

Passing to the limit as $n \to +\infty$, we get $\Phi(x, 0) - \Phi(a, 0) \geq 0$ for all $x \in X$, thus $a \in \mathrm{dom}\,\Phi(\cdot, 0)$ is an optimal solution of the problem (P_Φ). $\qquad \square$

Remark 3.1 Let us notice that in the setting of separated locally convex spaces the implications $(ii) \Rightarrow (iii) \Rightarrow (i)$ in the theorems 3.1 and 3.2 hold also in the case the hypothesis of lower semicontinuity of Φ is removed.

Remark 3.2 Using the convention mentioned at the beginning of the section, the above results can be reformulated as follows. Under the hypotheses of Theorem 3.1 the following assertions are equivalent:

(i) $a \in \mathrm{dom}\,\Phi(\cdot, 0)$ is an optimal solution of the problem (P_Φ);

(ii) there exist sequences $\{\varepsilon_n\} \downarrow 0$ and $(x_n^*, y_n^*) \in \partial_{\varepsilon_n} \Phi(a, 0)$ such that $x_n^* \to 0$ $(n \to +\infty)$;

(iii) there exist sequences $(x_n, y_n) \in \mathrm{dom}\, \Phi$ and $(x_n^*, y_n^*) \in \partial \Phi(x_n, y_n)$ such that

$$x_n^* \to 0,\ x_n \to a,\ y_n \to 0\ (n \to +\infty) \text{ and}$$

$$\Phi(x_n, y_n) - \langle y_n^*, y_n \rangle - \Phi(a, 0) \to 0\ (n \to +\infty).$$

Remark 3.3 Let us notice that a refined version of the above sequential characterizations expressed by means of the (convex) subdifferential can be given. To this aim one has to use an idea due to THIBAULT (cf. [137]) which we present in the following. Under the hypotheses of Theorem 1.1, applying this result to the indicator function of epi f, one obtains that for every $\varepsilon > 0$ and every $x^* \in \partial_\varepsilon f(a)$, there exist $(x_\varepsilon, r_\varepsilon) \in \mathrm{epi}\, f$ and $(x_\varepsilon^*, -s_\varepsilon) \in N_{\mathrm{epi}\, f}(x_\varepsilon, r_\varepsilon)$ such that

$$\|(x_\varepsilon, r_\varepsilon) - (a, f(a))\| \leq \sqrt{\varepsilon},\ \|(x_\varepsilon^*, -s_\varepsilon) - (x^*, -1)\| \leq \sqrt{\varepsilon}$$

and

$$|\langle x_\varepsilon^*, x_\varepsilon - a \rangle - s_\varepsilon(r_\varepsilon - f(a))| \leq 2\varepsilon.$$

This yields that

$$\|x_\varepsilon - a\| \leq \sqrt{\varepsilon},\ \|x_\varepsilon^* - x^*\|_* \leq \sqrt{\varepsilon},\ f(x_\varepsilon) - f(a) \leq \sqrt{\varepsilon},$$

$$|s_\varepsilon - 1| \leq \sqrt{\varepsilon} \text{ and } |\langle x_\varepsilon^*, x_\varepsilon - a \rangle| \leq 3\varepsilon + \sqrt{\varepsilon}.$$

Considering a sequence $\{\varepsilon_n\} \downarrow 0$ (for which we can assume without loss of generality that $\varepsilon_n < 1$ for all $n \in \mathbb{N}$) and $x_{\varepsilon_n}^* \in \partial_{\varepsilon_n} f(a)$, by defining $u_{\varepsilon_n}^* := (1/s_{\varepsilon_n})x_{\varepsilon_n}^*$, we obtain a family $(x_{\varepsilon_n}, u_{\varepsilon_n}^*)$ fulfilling $u_{\varepsilon_n}^* \in \partial f(x_{\varepsilon_n})$ for all $n \in \mathbb{N}$. Using that f is lower semicontinuous we further get

$$u_{\varepsilon_n}^* \xrightarrow{\|\cdot\|_*} x^*,\ x_{\varepsilon_n} \to a,\ f(x_{\varepsilon_n}) \to f(a) \text{ and } \langle u_{\varepsilon_n}^*, x_{\varepsilon_n} - a \rangle \to 0\ (n \to +\infty).$$

Let us mention that the conclusion above can be obtained also by applying [106, Proposition 1.1].

Employing the facts already described one can refine the results in the conclusion of Theorem 3.2. Considering the same hypotheses the following statements are equivalent:

(i) $a \in \mathrm{dom}\, \Phi(\cdot, 0)$ is an optimal solution of the problem (P_Φ);

(ii) there exist sequences $(x_n, y_n) \in \mathrm{dom}\, \Phi$ and $(x_n^*, y_n^*) \in \partial \Phi(x_n, y_n)$ such that

$$x_n^* \to 0,\ x_n \to a,\ y_n \to 0,\ \langle y_n^*, y_n \rangle \to 0\ (n \to +\infty) \text{ and}$$

$$\Phi(x_n, y_n) - \Phi(a, 0) \to 0\ (n \to +\infty).$$

However, we work in the following with the conditions given in the theorems 3.1 and 3.2, since their different particularizations deliver us several results from the literature dealing with sequential optimality conditions.

3.2   Sequential generalizations of the Pshenichnyi-Rockafellar Lemma

In this section we particularize the general sequential optimality conditions introduced above to the optimization problem

$$(P_F^A) \quad \inf_{x \in X} \{f(x) + (g \circ A)(x)\},$$

where $(X, \|\cdot\|)$ is a reflexive Banach space, $(Y, \|\cdot\|)$ is a Banach space, $f : X \to \overline{\mathbb{R}}$ and $g : Y \to \overline{\mathbb{R}}$ are proper, convex and lower semicontinuous functions and $A : X \to Y$ is a continuous linear mapping such that $A(\text{dom } f) \cap \text{dom } g \neq \emptyset$. To this end we define the perturbation function $\Phi_A : X \times Y \to \overline{\mathbb{R}}$ by $\Phi_A(x, y) = f(x) + g(Ax + y)$ for all $(x, y) \in X \times Y$. A simple computation shows that $\Phi_A^*(x^*, y^*) = f^*(x^* - A^*y^*) + g^*(y^*)$ for all $(x^*, y^*) \in X^* \times Y^*$. Let us prove first the following lemma.

Lemma 3.2 *Let* $(x^*, y^*) \in X^* \times Y^*$, $a \in \text{dom } f \cap A^{-1}(\text{dom } g)$ *and* $\varepsilon \geq 0$ *be fixed. The following statements are true:*

(i) *if* $(x^*, y^*) \in \partial_\varepsilon \Phi_A(a, 0)$, *then* $x^* - A^*y^* \in \partial_\varepsilon f(a)$ *and* $y^* \in \partial_\varepsilon g(Aa)$;

(ii) *if* $x^* - A^*y^* \in \partial_\varepsilon f(a)$ *and* $y^* \in \partial_\varepsilon g(Aa)$, *then* $(x^*, y^*) \in \partial_{2\varepsilon} \Phi_A(a, 0)$.

Proof. The pair (x^*, y^*) belongs to $\partial_\varepsilon \Phi_A(a, 0)$ if and only if $\Phi_A(a, 0) + \Phi_A^*(x^*, y^*) \leq \langle x^*, a \rangle + \varepsilon$, which is equivalent to $f(a) + g(Aa) + f^*(x^* - A^*y^*) + g^*(y^*) \leq \langle x^*, a \rangle + \varepsilon$.

(i) If $(x^*, y^*) \in \partial_\varepsilon \Phi_A(a, 0)$, then $f(a) + g(Aa) + f^*(x^* - A^*y^*) + g^*(y^*) \leq \langle x^*, a \rangle + \varepsilon$. Let us suppose that $x^* - A^*y^* \notin \partial_\varepsilon f(a)$. Then $f(a) + f^*(x^* - A^*y^*) > \langle x^* - A^*y^*, a \rangle + \varepsilon$. By the Young-Fenchel inequality we have $g(Aa) + g^*(y^*) \geq \langle y^*, Aa \rangle$. Adding the last two inequalities we obtain $f(a) + g(Aa) + f^*(x^* - A^*y^*) + g^*(y^*) > \langle x^*, a \rangle + \varepsilon$, which is a contradiction. Hence $x^* - A^*y^* \in \partial_\varepsilon f(a)$ and similarly we get $y^* \in \partial_\varepsilon g(Aa)$.

(ii) As $x^* - A^*y^* \in \partial_\varepsilon f(a)$ and $y^* \in \partial_\varepsilon g(Aa)$, we obtain $f(a) + f^*(x^* - A^*y^*) \leq \langle x^* - A^*y^*, a \rangle + \varepsilon$ and $g(Aa) + g^*(y^*) \leq \langle y^*, Aa \rangle + \varepsilon$. The conclusion follows by adding these two inequalities.   □

Theorem 3.3 *Let* $A : X \to Y$ *be a continuous linear mapping,* $f : X \to \overline{\mathbb{R}}$ *and* $g : Y \to \overline{\mathbb{R}}$ *be proper, convex and lower semicontinuous functions such that* $A(\text{dom } f) \cap \text{dom } g \neq \emptyset$. *Then* $a \in \text{dom } f \cap A^{-1}(\text{dom } g)$ *is an optimal solution of the problem* (P_F^A) *if and only if*

$$\exists \{\varepsilon_n\} \downarrow 0, \exists x_n^* \in \partial_{\varepsilon_n} f(a), \exists y_n^* \in \partial_{\varepsilon_n} g(Aa) \text{ such that } x_n^* + A^*y_n^* \to 0 \ (n \to +\infty). \tag{3. 1}$$

Proof. The element $a \in \text{dom } f \cap A^{-1}(\text{dom } g)$ is an optimal solution of the problem (P_F^A) if and only if a is an optimal solution of (P_{Φ_A}), which is equivalent to (cf. Theorem 3.1)

$$\exists \{\varepsilon_n\} \downarrow 0, \exists (x_n^*, y_n^*) \in \partial_{\varepsilon_n} \Phi_A(a, 0) \text{ such that } x_n^* \to 0 \ (n \to +\infty). \tag{3. 2}$$

We prove that the conditions (3. 1) and (3. 2) are equivalent.

"(3. 2)⇒(3. 1)" Assume that there exist $\{\overline{\varepsilon_n}\} \downarrow 0$ and $(\overline{x_n^*}, \overline{y_n^*}) \in \partial_{\overline{\varepsilon_n}} \Phi_A(a, 0)$ such that $\overline{x_n^*} \to 0 \ (n \to +\infty)$. According to Lemma 3.2(i), $\overline{x_n^*} - A^*\overline{y_n^*} \in \partial_{\overline{\varepsilon_n}} f(a)$ and $\overline{y_n^*} \in \partial_{\overline{\varepsilon_n}} g(Aa)$. By choosing $\varepsilon_n := \overline{\varepsilon_n}$, $x_n^* := \overline{x_n^*} - A^*\overline{y_n^*}$ and $y_n^* := \overline{y_n^*}$, we see that (3. 1) is fulfilled.

"(3. 1) ⇒(3. 2)" Assume that there exist $\{\overline{\varepsilon_n}\} \downarrow 0$, $\overline{x_n^*} \in \partial_{\overline{\varepsilon_n}} f(a)$ and $\overline{y_n^*} \in \partial_{\overline{\varepsilon_n}} g(Aa)$ such that $\overline{x_n^*} + A^*\overline{y_n^*} \to 0 \ (n \to +\infty)$. Take $\varepsilon_n := 2\overline{\varepsilon_n}$, $x_n^* := \overline{x_n^*} + A^*\overline{y_n^*}$ and $y_n^* := \overline{y_n^*}$. Then $x_n^* - A^*y_n^* = \overline{x_n^*} \in \partial_{\overline{\varepsilon_n}} f(a)$ and $y_n^* = \overline{y_n^*} \in \partial_{\overline{\varepsilon_n}} g(Aa)$, hence by Lemma 3.2(ii) we have $(x_n^*, y_n^*) \in \partial_{\varepsilon_n} \Phi_A(a, 0)$. Moreover, $x_n^* = \overline{x_n^*} + A^*\overline{y_n^*} \to 0 \ (n \to +\infty)$, so (3. 2) is fulfilled.   □

Remark 3.4 One can notice that the above characterization of the optimal solutions of the problem (P_F^A) can be also obtained as a consequence of the so-called *Hiriart-Urruty and Phelps Formula* (see [139, Proposition 1]).

Next we derive from Theorem 3.2 a sequential optimality condition for the problem (P_F^A) involving only the (convex) subdifferentials of the functions f and g.

Theorem 3.4 *Let* $A : X \to Y$ *be a continuous linear mapping,* $f : X \to \overline{\mathbb{R}}$ *and* $g : Y \to \overline{\mathbb{R}}$ *be proper, convex and lower semicontinuous functions such that* $A(\operatorname{dom} f) \cap \operatorname{dom} g \neq \emptyset$. *Then* $a \in \operatorname{dom} f \cap A^{-1}(\operatorname{dom} g)$ *is an optimal solution of the problem* (P_F^A) *if and only if*

$$
\begin{cases}
\exists (x_n, y_n) \in \operatorname{dom} f \times \operatorname{dom} g, \exists x_n^* \in \partial f(x_n), \exists y_n^* \in \partial g(y_n) \text{ such that} \\
x_n^* + A^* y_n^* \to 0, \ x_n \to a, \ y_n \to Aa \ (n \to +\infty), \\
f(x_n) - \langle x_n^*, x_n - a \rangle - f(a) \to 0, \ (n \to +\infty) \text{ and} \\
g(y_n) - \langle y_n^*, y_n - Aa \rangle - g(Aa) \to 0 \ (n \to +\infty).
\end{cases}
\tag{3.3}
$$

Proof. Applying Theorem 3.2, we get that a is an optimal solution of the problem (P_F^A) if and only if $\exists (x_n, y_n) \in X \times Y$, $x_n \in \operatorname{dom} f$, $Ax_n + y_n \in \operatorname{dom} g$, $\exists (x_n^*, y_n^*) \in \partial \Phi_A(x_n, y_n)$ such that $x_n^* \to 0, x_n \to a, y_n \to 0$ and $\Phi(x_n, y_n) - \langle y_n^*, y_n \rangle - \Phi(a, 0) \to 0$ $(n \to +\infty)$. The last condition is equivalent to

$$
f(x_n) + g(Ax_n + y_n) - \langle y_n^*, y_n \rangle - f(a) - g(Aa) \to 0 \ (n \to +\infty).
$$

For all $n \in \mathbb{N}$ we have $(x_n^*, y_n^*) \in \partial \Phi_A(x_n, y_n)$ if and only if $\Phi_A(x_n, y_n) + \Phi_A^*(x_n^*, y_n^*) = \langle x_n^*, x_n \rangle + \langle y_n^*, y_n \rangle \Leftrightarrow f(x_n) + g(Ax_n + y_n) + f^*(x_n^* - A^* y_n^*) + g^*(y_n^*) = \langle x_n^*, x_n \rangle + \langle y_n^*, y_n \rangle$. Using the Young-Fenchel inequality we obtain

$$
f(x_n) + f^*(x_n^* - A^* y_n^*) + g(Ax_n + y_n) + g^*(y_n^*) \geq \langle x_n^* - A^* y_n^*, x_n \rangle + \langle y_n^*, Ax_n + y_n \rangle
$$

$$
= \langle x_n^*, x_n \rangle + \langle y_n^*, y_n \rangle,
$$

hence $(x_n^*, y_n^*) \in \partial \Phi_A(x_n, y_n)$ if and only if $f(x_n) + f^*(x_n^* - A^* y_n^*) = \langle x_n^* - A^* y_n^*, x_n \rangle$ and $g(Ax_n + y_n) + g^*(y_n^*) = \langle y_n^*, Ax_n + y_n \rangle \Leftrightarrow x_n^* - A^* y_n^* \in \partial f(x_n)$ and $y_n^* \in \partial g(Ax_n + y_n)$. In this way we proved that $a \in \operatorname{dom} f \cap A^{-1}(\operatorname{dom} g)$ is an optimal solution of the problem (P_F^A) if and only if

$$
\begin{cases}
\exists (x_n, y_n) \in X \times Y, x_n \in \operatorname{dom} f, Ax_n + y_n \in \operatorname{dom} g, \\
\exists (x_n^*, y_n^*) \in X^* \times Y^*, x_n^* - A^* y_n^* \in \partial f(x_n), y_n^* \in \partial g(Ax_n + y_n) \text{ such that} \\
x_n^* \to 0, \ x_n \to a, \ y_n \to 0 \ (n \to +\infty) \text{ and} \\
f(x_n) + g(Ax_n + y_n) - \langle y_n^*, y_n \rangle - f(a) - g(Aa) \to 0 \ (n \to +\infty).
\end{cases}
\tag{3.4}
$$

Next we show that the conditions (3.3) and (3.4) are equivalent.

"(3.4)$\Rightarrow$(3.3)" Suppose that

$$
\begin{cases}
\exists (\overline{x}_n, \overline{y}_n) \in X \times Y, \overline{x}_n \in \operatorname{dom} f, A\overline{x}_n + \overline{y}_n \in \operatorname{dom} g, \\
\exists (\overline{x}_n^*, \overline{y}_n^*) \in X^* \times Y^*, \overline{x}_n^* - A^* \overline{y}_n^* \in \partial f(\overline{x}_n), \overline{y}_n^* \in \partial g(A\overline{x}_n + \overline{y}_n) \text{ such that} \\
\overline{x}_n^* \to 0, \ \overline{x}_n \to a, \ \overline{y}_n \to 0 \ (n \to +\infty) \text{ and} \\
f(\overline{x}_n) + g(A\overline{x}_n + \overline{y}_n) - \langle \overline{y}_n^*, \overline{y}_n \rangle - f(a) - g(Aa) \to 0 \ (n \to +\infty).
\end{cases}
$$

Take $x_n := \overline{x}_n, y_n := A\overline{x}_n + \overline{y}_n, x_n^* := \overline{x}_n^* - A^* \overline{y}_n^*$ and $y_n^* := \overline{y}_n^*$, for all $n \in \mathbb{N}$. Then $x_n \in \operatorname{dom} f, y_n \in \operatorname{dom} g, x_n^* \in \partial f(x_n), y_n^* \in \partial g(y_n), x_n^* + A^* y_n^* \to 0, x_n \to a$ and $y_n \to Aa$ $(n \to +\infty)$. Moreover,

$$
f(x_n) - \langle x_n^*, x_n - a \rangle - f(a) = f(\overline{x}_n) - \langle \overline{x}_n^* - A^* \overline{y}_n^*, \overline{x}_n - a \rangle - f(a)
$$

$$
= f(\overline{x}_n) + g(A\overline{x}_n + \overline{y}_n) - \langle \overline{x}_n^*, \overline{x}_n - a \rangle - \langle \overline{y}_n^*, \overline{y}_n \rangle - f(a) - g(Aa) - g(A\overline{x}_n + \overline{y}_n)
$$

$$+\langle A^* \overline{y_n^*}, \overline{x_n} - a \rangle + \langle \overline{y_n^*}, \overline{y_n} \rangle + g(Aa) = f(\overline{x_n}) + g(A\overline{x_n} + \overline{y_n}) - \langle \overline{x_n^*}, \overline{x_n} - a \rangle - \langle \overline{y_n^*}, \overline{y_n} \rangle$$

$$-f(a) - g(Aa) - g(y_n) + g(Aa) + \langle y_n^*, y_n - Aa \rangle.$$

Let us make the following notations: $a_n := f(x_n) - \langle x_n^*, x_n - a \rangle - f(a)$ and $b_n := g(Aa) - g(y_n) - \langle y_n^*, Aa - y_n \rangle$. We have $a_n - b_n = f(\overline{x_n}) + g(A\overline{x_n} + \overline{y_n}) - \langle \overline{y_n^*}, \overline{y_n} \rangle - f(a) - g(Aa) - \langle \overline{x_n^*}, \overline{x_n} - a \rangle \to 0$ $(n \to +\infty)$. Since $x_n^* \in \partial f(x_n)$ we have $f(x) - f(x_n) \geq \langle x_n^*, x - x_n \rangle$ for all $x \in X$. For $x = a$ in the previous inequality we get $a_n = f(x_n) - \langle x_n^*, x_n - a \rangle - f(a) \leq 0$. Similarly, from $y_n^* \in \partial g(y_n)$ we have $b_n = g(Aa) - g(y_n) - \langle y_n^*, Aa - y_n \rangle \geq 0$. Thus $a_n \leq 0 \leq b_n$ and $a_n - b_n \to 0$ $(n \to +\infty)$. As in this case one must have that $a_n \to 0$ and $b_n \to 0$ $(n \to \infty)$, (3. 3) is fulfilled.

"(3. 3)$\Rightarrow$(3. 4)" Assume now that (3. 3) holds, namely

$$\begin{cases} \exists (\overline{x_n}, \overline{y_n}) \in \mathrm{dom}\, f \times \mathrm{dom}\, g, \exists \overline{x_n^*} \in \partial f(\overline{x_n}), \exists \overline{y_n^*} \in \partial g(\overline{y_n}) \text{ such that} \\ \overline{x_n^*} + A^* \overline{y_n^*} \to 0, \; \overline{x_n} \to a, \; \overline{y_n} \to Aa \; (n \to +\infty), \\ f(\overline{x_n}) - \langle \overline{x_n^*}, \overline{x_n} - a \rangle - f(a) \to 0 \; (n \to +\infty) \text{ and} \\ g(\overline{y_n}) - \langle \overline{y_n^*}, \overline{y_n} - Aa \rangle - g(Aa) \to 0 \; (n \to +\infty). \end{cases}$$

For all $n \in \mathbb{N}$ take $x_n := \overline{x_n}, y_n := \overline{y_n} - A\overline{x_n}, y_n^* := \overline{y_n^*}$ and $x_n^* := \overline{x_n^*} + A^* \overline{y_n^*}$. Then $x_n \in \mathrm{dom}\, f, Ax_n + y_n \in \mathrm{dom}\, g, x_n^* - A^* y_n^* \in \partial f(x_n), y_n^* \in \partial g(Ax_n + y_n), x_n^* \to 0, x_n \to a$ and $y_n \to 0$ $(n \to +\infty)$. Moreover,

$$f(x_n) + g(Ax_n + y_n) - \langle y_n^*, y_n \rangle - f(a) - g(Aa) = f(\overline{x_n}) + g(\overline{y_n}) - \langle \overline{y_n^*}, \overline{y_n} - A\overline{x_n} \rangle$$

$$-f(a) - g(Aa) = f(\overline{x_n}) - \langle \overline{x_n^*}, \overline{x_n} - a \rangle - f(a) + g(\overline{y_n}) - \langle \overline{y_n^*}, \overline{y_n} - Aa \rangle - g(Aa)$$

$$+\langle \overline{x_n^*}, \overline{x_n} - a \rangle + \langle \overline{y_n^*}, -Aa + A\overline{x_n} \rangle = f(\overline{x_n}) - \langle \overline{x_n^*}, \overline{x_n} - a \rangle - f(a)$$

$$+g(\overline{y_n}) - \langle \overline{y_n^*}, \overline{y_n} - Aa \rangle - g(Aa) + \langle \overline{x_n^*} + A^* \overline{y_n^*}, \overline{x_n} - a \rangle \to 0 \; (n \to +\infty),$$

hence (3. 4) is fulfilled. $\qquad\square$

Remark 3.5 Similar characterizations of the optimal solutions of the problem (P_F^A) have been given by THIBAULT in [139] and as an application they have been used to provide a new proof of the well known fact that the subdifferential of a proper, convex and lower semicontinuous function is a maximal monotone operator (cf. [124]).

If we take $Y = X$ (X is a reflexive Banach space) and $A = \mathrm{id}_X$ in the above theorems we obtain the following sequential optimality conditions concerning the convex optimization problem

$$(P_F) \quad \inf_{x \in X} \{f(x) + g(x)\}.$$

They are presented in the following as two corollaries.

Corollary 3.1 Let $f, g : X \to \overline{\mathbb{R}}$ be proper, convex and lower semicontinuous functions such that $\mathrm{dom}\, f \cap \mathrm{dom}\, g \neq \emptyset$. Then $a \in \mathrm{dom}\, f \cap \mathrm{dom}\, g$ is an optimal solution of the problem (P_F) if and only if

$$\exists \{\varepsilon_n\} \downarrow 0, \exists x_n^* \in \partial_{\varepsilon_n} f(a), \exists y_n^* \in \partial_{\varepsilon_n} g(a) \text{ such that } x_n^* + y_n^* \to 0 \; (n \to +\infty).$$

Corollary 3.2 Let $f, g : X \to \overline{\mathbb{R}}$ be proper, convex and lower semicontinuous functions such that $\mathrm{dom}\, f \cap \mathrm{dom}\, g \neq \emptyset$. Then $a \in \mathrm{dom}\, f \cap \mathrm{dom}\, g$ is an optimal solution of the problem (P_F) if and only if

$$\begin{cases} \exists (x_n, y_n) \in \mathrm{dom}\, f \times \mathrm{dom}\, g, \exists x_n^* \in \partial f(x_n), \exists y_n^* \in \partial g(y_n) \text{ such that} \\ x_n^* + y_n^* \to 0, \; x_n \to a, \; y_n \to a \; (n \to +\infty), \\ f(x_n) - \langle x_n^*, x_n - a \rangle - f(a) \to 0 \; (n \to +\infty) \text{ and} \\ g(y_n) - \langle y_n^*, y_n - a \rangle - g(a) \to 0 \; (n \to +\infty). \end{cases}$$

Taking $g := \delta_D$ in the previous corollaries, where $D \subseteq X$ is a non-empty, closed and convex set, we obtain the following sequential optimality conditions regarding the convex optimization problem

$$(P_F^0) \quad \inf_{x \in D} f(x).$$

Corollary 3.3 Let $f : X \to \overline{\mathbb{R}}$ be a proper, convex and lower semicontinuous function and $D \subseteq X$ a closed and convex set such that $D \cap \operatorname{dom} f \neq \emptyset$. Then $a \in D \cap \operatorname{dom} f$ is an optimal solution of the problem (P_F^0) if and only if

$$\exists \{\varepsilon_n\} \downarrow 0, \exists x_n^* \in \partial_{\varepsilon_n} f(a), \exists y_n^* \in N_D^{\varepsilon_n}(a) \text{ such that } x_n^* + y_n^* \to 0 \ (n \to +\infty).$$

Corollary 3.4 Let $f : X \to \overline{\mathbb{R}}$ be a proper, convex and lower semicontinuous function and $D \subseteq X$ a closed and convex set such that $D \cap \operatorname{dom} f \neq \emptyset$. Then $a \in D \cap \operatorname{dom} f$ is an optimal solution of the problem (P_F^0) if and only if

$$\begin{cases} \exists (x_n, y_n) \in \operatorname{dom} f \times D, \exists x_n^* \in \partial f(x_n), \exists y_n^* \in N_D(y_n) \text{ such that} \\ x_n^* + y_n^* \to 0, \ x_n \to a, \ y_n \to a \ (n \to +\infty), \\ f(x_n) - \langle x_n^*, x_n - a \rangle - f(a) \to 0 \ (n \to +\infty) \text{ and} \\ \langle y_n^*, y_n - a \rangle \to 0 \ (n \to +\infty). \end{cases}$$

Remark 3.6 As shown in this section, the last two results are obtained as particular cases of the main results of this chapter, theorems 3.1 and 3.2. Moreover, corollaries 3.3 and 3.4 can be seen as sequential generalizations of the well-known Pshenichnyi-Rockafellar Lemma, improving in the same time the results of JEYAKUMAR AND WU (see [86, Theorem 3.3 and Corollary 3.5]). One can notice that in our case the convergence on X^* can be considered both in the weak* and strong topology, while in [86] only the weak* topology is taken.

Let us give in the following an example to show that even in the situation when the celebrated Pshenichnyi-Rockafellar Lemma fails, its sequential form given in Corollary 3.4 is applicable (for other examples illustrating the advantages of having sequential optimality conditions we refer to [86]).

Example 3.1 Let be $X = \mathbb{R}^2$, $D = -\mathbb{R}_+^2$ and the function $f : \mathbb{R}^2 \to \overline{\mathbb{R}}$ be defined by

$$f(x, y) = \begin{cases} x^2 - \sqrt{y}, & \text{if } y \geq 0, \\ +\infty, & \text{otherwise.} \end{cases}$$

Obviously $a = (0, 0) \in (-\infty, 0] \times \{0\} = D \cap \operatorname{dom} f$ is the unique optimal solution of the problem (P_F^0). Since $\partial f(a) = \emptyset$, the Pshenichnyi-Rockafellar Lemma cannot be applied. For all $n \in \mathbb{N}$ take $x_n = (0, 1/n) \in \operatorname{dom} f$, $y_n = (0, 0) \in D$, $x_n^* = (0, -\sqrt{n}/2) \in \partial f(x_n)$ and $y_n^* = (0, \sqrt{n}/2) \in \mathbb{R}_+^2 = N_D(y_n)$. We have $x_n \to a \ (n \to +\infty)$, $y_n = a$, $x_n^* + y_n^* = 0$, $f(x_n) - \langle x_n^*, x_n - a \rangle - f(a) = -\sqrt{1/n} + (1/2)(1/\sqrt{n}) = -1/(2\sqrt{n}) \to 0 \ (n \to +\infty)$ and $\langle y_n^*, y_n - a \rangle = 0$, hence the sequential optimality conditions in Corollary 3.4 are fulfilled.

3.3 Sequential optimality conditions for the problem with geometric and cone constraints

Let X be a reflexive Banach space, Y a Banach space and $C \subseteq Y$ a non-empty convex cone inducing a partial ordering on Y. In this section we deal with the convex optimization problem with geometric and cone constraints

$$(P_L) \quad \inf_{\substack{x \in S \\ g(x) \in -C}} f(x),$$

where $S \cap g^{-1}(-C) \cap \text{dom} f \neq \emptyset$, S is a closed convex subset of X, $f : X \to \overline{\mathbb{R}}$ is a proper, convex and lower semicontinuous function and $g : X \to Y^\bullet$ is a C-convex vector-valued function.

3.3.1   The case g is continuous

Additionally to the assumptions made above we suppose in this subsection that the cone C is closed and $g : X \to Y$ is continuous. We derive a sequential form of the Lagrange multiplier rule for (P_L) by applying Theorem 3.2 to the following perturbation function

$$\Phi_1^C : X \times X \times Y \to \overline{\mathbb{R}}, \ \Phi_1^C(x, p, q) = \begin{cases} f(x), & \text{if } x + p \in S \text{ and } g(x) \in q - C, \\ +\infty, & \text{otherwise.} \end{cases}$$

The conjugate of Φ_1^C is $(\Phi_1^C)^* : X^* \times X^* \times Y^* \to \overline{\mathbb{R}}$,

$$(\Phi_1^C)^*(x^*, p^*, q^*) = \sup_{\substack{(x,p,q) \in X \times X \times Y \\ x+p \in S \\ g(x) \in q-C}} \{\langle x^*, x \rangle + \langle p^*, p \rangle + \langle q^*, q \rangle - f(x)\}.$$

In order to compute $(\Phi_1^C)^*$ we introduce new variables z and y by $z := x + p$ and $q - g(x) := y$, respectively. It follows

$$(\Phi_1^C)^*(x^*, p^*, q^*) = \sup_{(x,z,y) \in X \times S \times C} \{\langle x^*, x \rangle + \langle p^*, z - x \rangle + \langle q^*, y + g(x) \rangle - f(x)\},$$

and, as the three variables are separated, we get $(\Phi_1^C)^*(x^*, p^*, q^*) = \sup_{z \in S} \langle p^*, z \rangle +$ $\sup_{x \in X} \{\langle x^* - p^*, x \rangle + \langle q^*, g(x) \rangle - f(x)\} + \sup_{y \in C} \langle q^*, y \rangle$. We obtain the following formula

$$(\Phi_1^C)^*(x^*, p^*, q^*) = \begin{cases} \delta_S^*(p^*) + \sup_{x \in X} \{\langle x^* - p^*, x \rangle + \langle q^*, g(x) \rangle - f(x)\}, & \text{if } q^* \in -C^*, \\ +\infty, & \text{otherwise.} \end{cases}$$

A direct application of Theorem 3.2 yields the following result.

Theorem 3.5 *The element $a \in S \cap g^{-1}(-C) \cap \text{dom} f$ is an optimal solution of the problem (P_L) if and only if*

$$\begin{cases} \exists (x_n, \omega_n, t_n) \in \text{dom} f \times S \times (-C), \exists (u_n^*, v_n^*, \omega_n^*, q_n^*) \in X^* \times X^* \times X^* \times C^*, \\ u_n^* \in \partial f(x_n), v_n^* \in \partial(q_n^* g)(x_n), \omega_n^* \in N_S(\omega_n), \langle q_n^*, t_n \rangle = 0 \ \forall n \in \mathbb{N}, \\ u_n^* + v_n^* + \omega_n^* \to 0, \ \omega_n \to a, \ x_n \to a, \ t_n \to g(a) \ (n \to +\infty) \ and \\ f(x_n) - f(a) + \langle q_n^*, g(x_n) \rangle - \langle \omega_n^*, \omega_n - x_n \rangle \to 0 \ (n \to +\infty). \end{cases}$$

$$(3.5)$$

Proof. According to Theorem 3.2, the element $a \in S \cap g^{-1}(-C) \cap \text{dom} f$ is an optimal solution of the problem (P_L) if and only if there exist sequences $(x_n, p_n, q_n) \in \text{dom} \Phi_1^C$ and $(x_n^*, p_n^*, -q_n^*) \in \partial \Phi_1^C(x_n, p_n, q_n)$ such that

$$x_n^* \to 0, x_n \to a, (p_n, q_n) \to (0, 0) \ (n \to +\infty) \ and$$

$$\Phi_1^C(x_n, p_n, q_n) - \langle (p_n^*, -q_n^*), (p_n, q_n) \rangle - \Phi_1^C(a, 0, 0) \to 0 \ (n \to +\infty).$$

Since $(x_n, p_n, q_n) \in \text{dom} \Phi_1^C$, we get $x_n \in \text{dom} f, x_n + p_n \in S$ and $g(x_n) \in q_n - C$, for all $n \in \mathbb{N}$. We have $(x_n^*, p_n^*, -q_n^*) \in \partial \Phi_1^C(x_n, p_n, q_n)$ if and only if

$$\Phi_1^C(x_n, p_n, q_n) + (\Phi_1^C)^*(x_n^*, p_n^*, -q_n^*) = \langle x_n^*, x_n \rangle + \langle p_n^*, p_n \rangle + \langle -q_n^*, q_n \rangle$$

$$\Leftrightarrow f(x_n) + \delta_S^*(p_n^*) + (f + q_n^* g)^*(x_n^* - p_n^*) = \langle x_n^*, x_n \rangle + \langle p_n^*, p_n \rangle + \langle -q_n^*, q_n \rangle.$$

The previous relation holds if and only if

$$(f + q_n^* g)^*(x_n^* - p_n^*) + (f + q_n^* g)(x_n) - \langle x_n^* - p_n^*, x_n \rangle$$

$$+ \langle q_n^*, q_n - g(x_n) \rangle + \delta_S^*(p_n^*) - \langle p_n^*, x_n + p_n \rangle = 0 \; \forall n \in \mathbb{N}.$$

As $q_n - g(x_n) \in C$ and $q_n^* \in C^*$, we have $\langle q_n^*, q_n - g(x_n) \rangle \geq 0$ for all $n \in \mathbb{N}$. Moreover, the Young-Fenchel inequality yields

$$(f + q_n^* g)^*(x_n^* - p_n^*) + (f + q_n^* g)(x_n) - \langle x_n^* - p_n^*, x_n \rangle \geq 0$$

and

$$\delta_S^*(p_n^*) - \langle p_n^*, x_n + p_n \rangle \geq 0,$$

hence $(x_n^*, p_n^*, -q_n^*) \in \partial \Phi_1^C(x_n, p_n, q_n)$ if and only if $x_n^* - p_n^* \in \partial (f + q_n^* g)(x_n), p_n^* \in \partial \delta_S(x_n + p_n) = N_S(x_n + p_n)$ and $\langle q_n^*, q_n - g(x_n) \rangle = 0$ for all $n \in \mathbb{N}$. The relation $\Phi_1^C(x_n, p_n, q_n) - \langle (p_n^*, -q_n^*), (p_n, q_n) \rangle - \Phi_1^C(a, 0, 0) \rightarrow 0 \; (n \rightarrow +\infty)$ is equivalent to $f(x_n) - \langle p_n^*, p_n \rangle + \langle q_n^*, q_n \rangle - f(a) \rightarrow 0 \; (n \rightarrow +\infty)$. Hence the element $a \in S \cap g^{-1}(-C) \cap \text{dom} \, f$ is an optimal solution of the problem (P_L) if and only if

$$\begin{cases} \exists (x_n, p_n, q_n) \in \text{dom} \, f \times X \times Y, x_n + p_n \in S, g(x_n) - q_n \in -C, \\ \exists (x_n^*, p_n^*, q_n^*) \in X^* \times X^* \times C^* \text{ such that} \\ x_n^* - p_n^* \in \partial (f + q_n^* g)(x_n), p_n^* \in N_S(x_n + p_n), \langle q_n^*, q_n - g(x_n) \rangle = 0 \; \forall n \in \mathbb{N}, \\ x_n^* \rightarrow 0, \; x_n \rightarrow a, \; p_n \rightarrow 0, \; q_n \rightarrow 0 \; (n \rightarrow +\infty) \text{ and} \\ f(x_n) - f(a) + \langle q_n^*, q_n \rangle - \langle p_n^*, p_n \rangle \rightarrow 0 \; (n \rightarrow +\infty). \end{cases}$$

$$(3. \; 6)$$

Introducing the new variables $t_n, \omega_n, \overline{u_n^*}$ and ω_n^* instead of q_n, p_n, x_n^* and p_n^*, by $t_n := g(x_n) - q_n, \omega_n := p_n + x_n, \overline{u_n^*} := x_n^* - p_n^*$ and $\omega_n^* := p_n^*$ for all $n \in \mathbb{N}$, respectively, the condition (3. 6) can be reformulated as follows

$$\begin{cases} \exists (x_n, \omega_n, t_n) \in \text{dom} \, f \times S \times (-C), \exists (\overline{u_n^*}, \omega_n^*, q_n^*) \in X^* \times X^* \times C^*, \\ \overline{u_n^*} \in \partial (f + q_n^* g)(x_n), \omega_n^* \in N_S(\omega_n), \langle q_n^*, t_n \rangle = 0 \; \forall n \in \mathbb{N}, \\ \overline{u_n^*} + \omega_n^* \rightarrow 0, \; \omega_n \rightarrow a, \; x_n \rightarrow a, \; t_n \rightarrow g(a) \; (n \rightarrow +\infty) \text{ and} \\ f(x_n) - f(a) + \langle q_n^*, g(x_n) \rangle - \langle \omega_n^*, \omega_n - x_n \rangle \rightarrow 0 \; (n \rightarrow +\infty). \end{cases}$$

$$(3. \; 7)$$

The function g being continuous, we obtain that the following subdifferential sum formula holds

$$\partial (f + q_n^* g)(x_n) = \partial f(x_n) + \partial (q_n^* g)(x_n)$$

(see [149, Theorem 2.8.7]). Thus $\overline{u_n^*} \in \partial (f + q_n^* g)(x_n)$ if and only if there exist $u_n^* \in \partial f(x_n)$ and $v_n^* \in \partial (q_n^* g)(x_n)$ such that $\overline{u_n^*} = u_n^* + v_n^*$ for all $n \in \mathbb{N}$ and the desired conclusion follows. $\qquad \square$

Remark 3.7 Let us introduce now the following real sequences: $l_n := f(x_n) - f(a) + \langle q_n^*, g(x_n) \rangle - \langle \omega_n^*, \omega_n - x_n \rangle$ (see Theorem 3.5), $l_n^1 := \langle q_n^*, t_n - g(a) \rangle + \langle \omega_n^*, \omega_n - a \rangle$ and $l_n^2 := f(x_n) - f(a) + \langle q_n^*, g(x_n) - g(a) \rangle + \langle \omega_n^*, x_n - a \rangle$ for all $n \in \mathbb{N}$. We prove that if the condition

$$\begin{cases} (x_n, \omega_n, t_n) \in \text{dom} \, f \times S \times (-C), (u_n^*, v_n^*, \omega_n^*, q_n^*) \in X^* \times X^* \times X^* \times C^*, \\ u_n^* \in \partial f(x_n), v_n^* \in \partial (q_n^* g)(x_n), \omega_n^* \in N_S(\omega_n), \langle q_n^*, t_n \rangle = 0 \; \forall n \in \mathbb{N} \text{ and} \\ u_n^* + v_n^* + \omega_n^* \rightarrow 0, x_n \rightarrow a \; (n \rightarrow +\infty), \end{cases}$$

$$(3. \; 8)$$

is satisfied, then we have

$$l_n \rightarrow 0 \; (n \rightarrow +\infty) \text{ if and only if } l_n^1 \rightarrow 0 \text{ and } l_n^2 \rightarrow 0 \; (n \rightarrow +\infty). \qquad (3. \; 9)$$

Indeed, if (3. 8) is fulfilled, then

$$l_n = l_n^2 - l_n^1 , \qquad (3.\ 10)$$

hence the sufficiency of relation (3. 9) is trivial (in fact for this implication we need only the fulfillment of $\langle q_n^*, t_n \rangle = 0$ for all $n \in \mathbb{N}$).

Assume now that $l_n \to 0$ $(n \to +\infty)$. Since $w_n^* \in N_S(w_n)$, we have $\langle w_n^*, a - w_n \rangle \leq 0$ and, as $q_n^* \in C^*$, we get

$$l_n^1 \geq 0 \ \forall n \in \mathbb{N}. \qquad (3.\ 11)$$

From $v_n^* \in \partial(q_n^* g)(x_n)$ we obtain the inequality $(q_n^* g)(a) - (q_n^* g)(x_n) \geq \langle v_n^*, a - x_n \rangle$, that is $\langle q_n^*, g(x_n) - g(a) \rangle \leq \langle v_n^*, x_n - a \rangle$. This inequality leads to $l_n^2 \leq f(x_n) - f(a) + \langle v_n^* + w_n^*, x_n - a \rangle$ for all $n \in \mathbb{N}$. Since $u_n^* \in \partial f(x_n)$ we have $f(a) - f(x_n) \geq \langle u_n^*, a - x_n \rangle$ for all $n \in \mathbb{N}$. Combining the last two inequalities we obtain $l_n^2 \leq \langle u_n^* + v_n^* + w_n^*, x_n - a \rangle$ for all $n \in \mathbb{N}$. This implies, using relation (3. 10) and inequality (3. 11), that

$$0 \leq l_n^1 = l_n^2 - l_n \leq \langle u_n^* + v_n^* + w_n^*, x_n - a \rangle - l_n \ \forall n \in \mathbb{N},$$

hence $l_n^1 \to 0$ $(n \to +\infty)$. From (3. 10) we obtain that $l_n^2 \to 0$ $(n \to +\infty)$.

By using the remarks made above we can state the following result.

Theorem 3.6 *The element $a \in S \cap g^{-1}(-C) \cap \operatorname{dom} f$ is an optimal solution of the problem (P_L) if and only if*

$$\begin{cases} \exists (x_n, w_n, t_n) \in \operatorname{dom} f \times S \times (-C), \exists (u_n^*, v_n^*, w_n^*, q_n^*) \in X^* \times X^* \times X^* \times C^*, \\ u_n^* \in \partial f(x_n), v_n^* \in \partial(q_n^* g)(x_n), w_n^* \in N_S(w_n), \langle q_n^*, t_n \rangle = 0 \ \forall n \in \mathbb{N}, \\ u_n^* + v_n^* + w_n^* \to 0, \ w_n \to a, \ x_n \to a, \ t_n \to g(a) \ (n \to +\infty), \\ \langle q_n^*, t_n - g(a) \rangle + \langle w_n^*, w_n - a \rangle \to 0 \ (n \to +\infty) \ and \\ f(x_n) - f(a) + \langle q_n^*, g(x_n) - g(a) \rangle + \langle w_n^*, x_n - a \rangle \to 0 \ (n \to +\infty). \end{cases}$$

$$(3.\ 12)$$

In order to show the applicability of the last statement, let us consider the following example.

Example 3.2 We work in the following setting: $X = Y = \mathbb{R}$, $S = C = \mathbb{R}_+$ and the functions $f : \mathbb{R} \to \overline{\mathbb{R}}$,

$$f(x) = \begin{cases} -\sqrt{x}, & \text{if } x \geq 0, \\ +\infty, & \text{otherwise,} \end{cases}$$

and $g : \mathbb{R} \to \mathbb{R}$, $g(x) = x$, for all $x \in \mathbb{R}$. Trivially, $a = 0$ is the unique optimal solution of the problem (P_L). For all $n \in \mathbb{N}$ take $x_n = 1/n \in \operatorname{dom} f$, $w_n = 0 \in S$, $t_n = 0 \in -C$, $u_n^* = -\sqrt{n}/2 \in \partial f(x_n)$, $q_n^* = \sqrt{n}/2 \in \mathbb{R}_+ = C^*$, $v_n^* = \sqrt{n}/2 \in \partial(q_n^* g)(x_n)$ and $w_n^* = 0 \in -\mathbb{R}_+ = N_S(w_n)$. It holds $f(x_n) - f(a) + \langle q_n^*, g(x_n) - g(a) \rangle + \langle w_n^*, x_n - a \rangle = f(x_n) + \langle q_n^*, g(x_n) \rangle = -\sqrt{1/n} + (\sqrt{n}/2)(1/n) = -1/(2\sqrt{n}) \to 0$ $(n \to +\infty)$ and one can easily see that all the conditions in Theorem 3.6 are fulfilled.

In case $\operatorname{dom} f = X$ and f is continuous we obtain from Theorem 3.6 the following corollary.

Corollary 3.5 *The element $a \in S \cap g^{-1}(-C) \cap \operatorname{dom} f$ is an optimal solution of the problem (P_L) if and only if*

$$\begin{cases} \exists (x_n, w_n, t_n) \in X \times S \times (-C), \exists (u_n^*, v_n^*, w_n^*, q_n^*) \in X^* \times X^* \times X^* \times C^*, \\ u_n^* \in \partial f(x_n), v_n^* \in \partial(q_n^* g)(x_n), w_n^* \in N_S(w_n), \langle q_n^*, t_n \rangle = 0 \ \forall n \in \mathbb{N}, \\ u_n^* + v_n^* + w_n^* \to 0, \ w_n \to a, \ x_n \to a, \ t_n \to g(a) \ (n \to +\infty), \\ \langle q_n^*, t_n - g(a) \rangle + \langle w_n^*, w_n - a \rangle \to 0 \ (n \to +\infty) \ and \\ \langle q_n^*, g(x_n) - g(a) \rangle + \langle w_n^*, x_n - a \rangle \to 0 \ (n \to +\infty). \end{cases}$$

$$(3.\ 13)$$

Remark 3.8 THIBAULT obtained the same characterization as in Corollary 3.5 under the additional hypotheses that C is a normal cone and Y is reflexive (cf. [138, Theorem 4.1]). Let us recall that $C \subseteq Y$ is normal if and only if the norm function (on Y) is C-increasing on C. Although in [138] it is not mentioned, the pair (ω_n, t_n) must belong to the set $S \times (-C)$ for all $n \in \mathbb{N}$ and if one looks carefully at the proof given by THIBAULT, one can see that this must be assumed also in [138, Theorem 4.1].

3.3.2 The case g is C-epi-closed

Let us consider the setting from the beginning of Section 3.3. Additionally we suppose that $g : X \to Y^\bullet$ is C-epi-closed. In the following we derive another sequential form of the Lagrange multiplier rule for the problem (P_L) by applying again Theorem 3.2, this time to the following perturbation function

$$\Phi_2^C : X \times X \times Y \to \overline{\mathbb{R}}, \ \Phi_2^C(x,p,q) = \begin{cases} f(x+p), & \text{if } x \in S \text{ and } g(x) \in q - C, \\ +\infty, & \text{otherwise.} \end{cases}$$

One can easily show that Φ_2^C is proper, convex and lower semicontinuous such that $(0,0) \in \mathrm{pr}_{X \times Y}(\mathrm{dom}\, \Phi_2^C)$. The conjugate of Φ_2^C is $(\Phi_2^C)^* : X^* \times X^* \times Y^* \to \overline{\mathbb{R}}$,

$$(\Phi_2^C)^*(x^*,p^*,q^*) = \begin{cases} f^*(p^*) + (-q^*g + \delta_S)^*(x^* - p^*), & \text{if } q^* \in -C^*, \\ +\infty, & \text{otherwise,} \end{cases}$$

as a straightforward calculation shows.

Theorem 3.7 *The element $a \in S \cap g^{-1}(-C) \cap \mathrm{dom}\, f$ is an optimal solution of the problem (P_L) if and only if*

$$\begin{cases} \exists (x_n, p_n, q_n) \in S \times \mathrm{dom}\, f \times Y, g(x_n) \leq_C q_n, \exists (u_n^*, v_n^*, q_n^*) \in X^* \times X^* \times C^*, \\ u_n^* \in \partial f(p_n), v_n^* \in \partial(q_n^*g + \delta_S)(x_n), \langle q_n^*, q_n - g(x_n) \rangle = 0 \ \forall n \in \mathbb{N}, \\ u_n^* + v_n^* \to 0, \ x_n \to a, \ p_n \to a, \ q_n \to 0 \ (n \to +\infty) \text{ and} \\ f(p_n) - \langle u_n^*, p_n - x_n \rangle + \langle q_n^*, q_n \rangle - f(a) \to 0 \ (n \to +\infty). \end{cases}$$

$$(3.\ 14)$$

Proof. According to Theorem 3.2, the element $a \in S \cap g^{-1}(-C) \cap \mathrm{dom}\, f$ is an optimal solution of the problem (P_L) if and only if there exist sequences $(x_n, p_n, q_n) \in \mathrm{dom}\, \Phi_2^C$, $(x_n^*, p_n^*, -q_n^*) \in \partial \Phi_2^C(x_n, p_n, q_n)$ such that

$$x_n^* \to 0, x_n \to a, (p_n, q_n) \to (0,0) \ (n \to +\infty) \text{ and}$$

$$\Phi_2^C(x_n, p_n, q_n) - \langle (p_n^*, -q_n^*), (p_n, q_n) \rangle - \Phi_2^C(a, 0, 0) \to 0 \ (n \to +\infty).$$

Since $(x_n, p_n, q_n) \in \mathrm{dom}\, \Phi_2^C$ we get $x_n \in S, x_n + p_n \in \mathrm{dom}\, f$ and $g(x_n) \leq_C q_n$ for all $n \in \mathbb{N}$. We have $(x_n^*, p_n^*, -q_n^*) \in \partial \Phi_2^C(x_n, p_n, q_n)$ if and only if

$$\Phi_2^C(x_n, p_n, q_n) + (\Phi_2^C)^*(x_n^*, p_n^*, -q_n^*) = \langle x_n^*, x_n \rangle + \langle p_n^*, p_n \rangle + \langle -q_n^*, q_n \rangle \Leftrightarrow$$

$$f(x_n + p_n) + f^*(p_n^*) + (q_n^*g + \delta_S)^*(x_n^* - p_n^*) = \langle x_n^*, x_n \rangle + \langle p_n^*, p_n \rangle + \langle -q_n^*, q_n \rangle,$$

where $q_n^* \in C^*$ for all $n \in \mathbb{N}$. As $q_n - g(x_n) \in C$ we obtain $\langle q_n^*, q_n - g(x_n) \rangle \geq 0$ for all $n \in \mathbb{N}$. Using this and the Young-Fenchel inequality we get $f(x_n + p_n) + f^*(p_n^*) + (q_n^*g + \delta_S)^*(x_n^* - p_n^*) \geq \langle p_n^*, x_n + p_n \rangle + \langle x_n^* - p_n^*, x_n \rangle - (q_n^*g + \delta_S)(x_n) = \langle x_n^*, x_n \rangle + \langle p_n^*, p_n \rangle + \langle -q_n^*, g(x_n) \rangle \geq \langle x_n^*, x_n \rangle + \langle p_n^*, p_n \rangle + \langle -q_n^*, q_n \rangle$. Hence $(x_n^*, p_n^*, -q_n^*) \in \partial \Phi_2^C(x_n, p_n, q_n)$ if and only if $p_n^* \in \partial f(x_n + p_n), x_n^* - p_n^* \in \partial(q_n^*g + \delta_S)(x_n)$ and

$\langle q_n^*, q_n - g(x_n) \rangle = 0$ for all $n \in \mathbb{N}$. As a consequence, we obtain that $a \in S \cap g^{-1}(-C) \cap \text{dom } f$ is an optimal solution of the problem (P_L) if and only if

$$
\begin{cases}
\exists (x_n, p_n, q_n) \in S \times X \times Y, x_n + p_n \in \text{dom } f, g(x_n) \leq_C q_n, \\
\exists (x_n^*, p_n^*, q_n^*) \in X^* \times X^* \times C^*, p_n^* \in \partial f(x_n + p_n), x_n^* - p_n^* \in \partial (q_n^* g + \delta_S)(x_n), \\
\langle q_n^*, q_n - g(x_n) \rangle = 0 \; \forall n \in \mathbb{N}, x_n^* \to 0, x_n \to a, p_n \to 0, q_n \to 0 \; (n \to +\infty) \text{ and} \\
f(x_n + p_n) - \langle p_n^*, p_n \rangle + \langle q_n^*, q_n \rangle - f(a) \to 0 \; (n \to +\infty).
\end{cases}
$$

$$(3.\ 15)$$

Introducing the new variables p_n', u_n^* and v_n^* instead of p_n, p_n^* and x_n^* by $p_n' := x_n + p_n, u_n^* := p_n^*$ and $v_n^* := x_n^* - p_n^*$ for all $n \in \mathbb{N}$, respectively, one can see that (3. 15) is equivalent to (3. 14) (again denoting p_n' by p_n for all $n \in \mathbb{N}$), which completes the proof. $\qquad \square$

For the special case when $S = X$, we obtain the following sequential characterization of the optimal solutions of the optimization problem

$$(P_L') \quad \inf_{g(x) \in -C} f(x).$$

We use this result in subsection 3.4.1 when deriving necessary and sufficient sequential optimality conditions for composed convex optimization problems.

Corollary 3.6 *The element $a \in g^{-1}(-C) \cap \text{dom } f$ is an optimal solution of the problem (P_L') if and only if*

$$
\begin{cases}
\exists (x_n, p_n, q_n) \in X \times \text{dom } f \times Y, g(x_n) \leq_C q_n, \exists (u_n^*, v_n^*, q_n^*) \in X^* \times X^* \times C^*, \\
u_n^* \in \partial f(p_n), v_n^* \in \partial (q_n^* g)(x_n), \langle q_n^*, q_n - g(x_n) \rangle = 0 \; \forall n \in \mathbb{N}, \\
u_n^* + v_n^* \to 0, \; x_n \to a, \; p_n \to a, \; q_n \to 0 \; (n \to +\infty) \text{ and} \\
f(p_n) - \langle u_n^*, p_n - x_n \rangle + \langle q_n^*, q_n \rangle - f(a) \to 0 \; (n \to +\infty).
\end{cases}
$$

$$(3.\ 16)$$

3.4 Sequential optimality conditions for composed convex optimization problems

Let us turn our attention to a more general optimization problem than the ones treated in the last two sections. We work in the following setting: X is a reflexive Banach space and Y is a Banach space partially ordered by the non-empty convex cone $C \subseteq Y$. The optimization problem considered in this section is

$$(P^{CC}) \quad \inf_{x \in X} \{ f(x) + (g \circ h)(x) \},$$

where $f : X \to \overline{\mathbb{R}}$ is proper, convex and lower semicontinuous, $h : X \to Y^\bullet$ is proper and C-convex and $g : Y^\bullet \to \overline{\mathbb{R}}$ is proper, convex and lower semicontinuous with $g(\infty_C) = +\infty$. We suppose also that $\text{dom } f \cap \text{dom } h \cap h^{-1}(\text{dom } g) \neq \emptyset$.

3.4.1 The case h is C-epi-closed

Throughout this subsection we assume additionally that Y is reflexive, h is C-epi-closed and g is C-increasing on $h(\text{dom } h) + C$. The problem (P^{CC}) is a convex optimization problem and for characterizing its optimal solutions the following sequential optimality condition can be derived from Corollary 3.6 (see Remark 3.12(b) for a discussion on the several reasons why we apply this method).

Theorem 3.8 *The element* $a \in \operatorname{dom} f \cap \operatorname{dom} h \cap h^{-1}(\operatorname{dom} g)$ *is an optimal solution of the problem* (P^{CC}) *if and only if*

$$
\left\{
\begin{array}{l}
\exists (x_n, p_n, q_n, q_n') \in X \times \operatorname{dom} f \times \operatorname{dom} g \times Y, h(x_n) \leq_C q_n', \\
\exists (u_n^*, e_n^*, u_n^{*\prime}, q_n^*), q_n^* \in C^*, u_n^* \in \partial f(p_n), q_n^* + e_n^* \in \partial g(q_n), \\
u_n^{*\prime} \in \partial(q_n^* h)(x_n), \langle q_n^*, q_n' - h(x_n) \rangle = 0 \ \forall n \in \mathbb{N}, \\
u_n^* + u_n^{*\prime} \to 0, \ e_n^* \to 0, \ x_n \to a, \ p_n \to a, \ q_n \to h(a), \ q_n' \to h(a) \ (n \to +\infty), \\
f(p_n) - \langle u_n^*, p_n - x_n \rangle + \langle q_n^*, h(x_n) - h(a) \rangle - f(a) \to 0 \ (n \to +\infty) \ and \\
g(q_n) - \langle q_n^*, q_n - h(a) \rangle - g(h(a)) \to 0 \ (n \to +\infty).
\end{array}
\right.
$$

$$(3.\ 17)$$

Proof. One can prove that $a \in \operatorname{dom} f \cap \operatorname{dom} h \cap h^{-1}(\operatorname{dom} g)$ is an optimal solution of the problem (P^{CC}) if and only if $(a, h(a))$ is an optimal solution of the problem

$$(P_L') \qquad \inf_{G(x,y) \in -C} F(x,y),$$

where $F : X \times Y \to \overline{\mathbb{R}}, F(x,y) = f(x) + g(y)$ and $G : X \times Y \to Y^\bullet, G(x,y) = h(x) - y$ for all $(x,y) \in X \times Y$. The hypotheses regarding the functions f, g and h imply that F is proper, convex and lower semicontinuous, while G is proper, C-convex and C-epi-closed. Applying Corollary 3.6 to the problem (P_L'), which is an optimization problem with cone constraints in $X \times Y$, we get that $a \in \operatorname{dom} f \cap \operatorname{dom} h \cap h^{-1}(\operatorname{dom} g)$ is an optimal solution of the problem (P^{CC}) if and only if

$$
\left\{
\begin{array}{l}
\exists (x_n, y_n, p_n, q_n, q_n') : (p_n, q_n) \in \operatorname{dom} F, G(x_n, y_n) \leq_C q_n', \\
\exists (u_n^*, v_n^*, u_n^{*\prime}, v_n^{*\prime}, q_n^*) : q_n^* \in C^*, (u_n^*, v_n^*) \in \partial F(p_n, q_n), \\
(u_n^{*\prime}, v_n^{*\prime}) \in \partial(q_n^* G)(x_n, y_n), \langle q_n^*, q_n' - G(x_n, y_n) \rangle = 0 \ \forall n \in \mathbb{N}, \\
(u_n^*, v_n^*) + (u_n^{*\prime}, v_n^{*\prime}) \to (0,0), (x_n, y_n) \to (a, h(a)), (p_n, q_n) \to (a, h(a)), \\
q_n' \to 0 \ \text{and} \ F(p_n, q_n) - \langle (u_n^*, v_n^*), (p_n, q_n) - (x_n, y_n) \rangle + \\
\langle q_n^*, q_n' \rangle - F(a, h(a)) \to 0 \ (n \to +\infty).
\end{array}
\right.
$$

$$(3.\ 18)$$

We have $\operatorname{dom} F = \operatorname{dom} f \times \operatorname{dom} g$, $F^*(x^*, y^*) = f^*(x^*) + g^*(y^*)$ and thus $(x^*, y^*) \in \partial F(x,y) \Leftrightarrow x^* \in \partial f(x)$ and $y^* \in \partial g(y)$, for $(x,y) \in X \times Y$ and $(x^*, y^*) \in X^* \times Y^*$. Further, for $\lambda \in C^*$ it holds

$$
(\lambda G)^*(x^*, y^*) = \left\{
\begin{array}{ll}
(\lambda h)^*(x^*), & \text{if } y^* + \lambda = 0, \\
+\infty, & \text{otherwise},
\end{array}
\right.
$$

and $(x^*, y^*) \in \partial(\lambda G)(x,y)$ if and only if $y^* + \lambda = 0$ and $x^* \in \partial(\lambda h)(x)$. Hence $a \in \operatorname{dom} f \cap \operatorname{dom} h \cap h^{-1}(\operatorname{dom} g)$ is an optimal solution of the problem (P^{CC}) if and only if

$$
\left\{
\begin{array}{l}
\exists (x_n, y_n, p_n, q_n, q_n') \in X \times Y \times \operatorname{dom} f \times \operatorname{dom} g \times Y : h(x_n) \leq_C y_n + q_n', \\
\exists (u_n^*, v_n^*, u_n^{*\prime}, q_n^*) : q_n^* \in C^*, u_n^* \in \partial f(p_n), v_n^* \in \partial g(q_n), u_n^{*\prime} \in \partial(q_n^* h)(x_n), \\
\langle q_n^*, q_n' + y_n - h(x_n) \rangle = 0 \ \forall n \in \mathbb{N}, u_n^* + u_n^{*\prime} \to 0, v_n^* - q_n^* \to 0, x_n \to a, p_n \to a, \\
y_n \to h(a), q_n \to h(a), q_n' \to 0 \ (n \to +\infty) \ \text{and} \ f(p_n) + g(q_n) - \langle u_n^*, p_n - x_n \rangle - \\
\langle v_n^*, q_n - y_n \rangle + \langle q_n^*, q_n' \rangle - f(a) - g(h(a)) \to 0 \ (n \to +\infty).
\end{array}
\right.
$$

$$(3.\ 19)$$

With the notations $q_n'' := y_n + q_n'$ and $e_n^* := v_n^* - q_n^*$ for all $n \in \mathbb{N}$, we obtain that (3. 19) is equivalent to

$$
\left\{
\begin{array}{l}
\exists (x_n, y_n, p_n, q_n, q_n'') \in X \times Y \times \operatorname{dom} f \times \operatorname{dom} g \times Y : h(x_n) \leq_C q_n'', \\
\exists (u_n^*, e_n^*, u_n^{*\prime}, q_n^*) : q_n^* \in C^*, u_n^* \in \partial f(p_n), q_n^* + e_n^* \in \partial g(q_n), u_n^{*\prime} \in \partial(q_n^* h)(x_n), \\
\langle q_n^*, q_n'' - h(x_n) \rangle = 0 \ \forall n \in \mathbb{N}, u_n^* + u_n^{*\prime} \to 0, e_n^* \to 0, x_n \to a, p_n \to a, y_n \to h(a), \\
q_n \to h(a), q_n'' \to h(a) \ (n \to +\infty) \ \text{and} \ f(p_n) + g(q_n) - \langle u_n^*, p_n - x_n \rangle - \\
\langle q_n^* + e_n^*, q_n - y_n \rangle + \langle q_n^*, q_n'' - y_n \rangle - f(a) - g(h(a)) \to 0 \ (n \to +\infty).
\end{array}
\right.
$$

$$(3.\ 20)$$

Since $\langle e_n^*, q_n - y_n \rangle \to 0$ $(n \to +\infty)$, we obtain that the element $a \in \mathrm{dom}\, f \cap \mathrm{dom}\, h \cap h^{-1}(\mathrm{dom}\, g)$ is an optimal solution of the problem (P^{CC}) if and only if

$$
\begin{cases}
\exists (x_n, y_n, p_n, q_n, q_n'') \in X \times Y \times \mathrm{dom}\, f \times \mathrm{dom}\, g \times Y : h(x_n) \leq_C q_n'', \\
\exists (u_n^*, e_n^*, u_n^{*\prime}, q_n^*) : q_n^* \in C^*, u_n^* \in \partial f(p_n), q_n^* + e_n^* \in \partial g(q_n), u_n^{*\prime} \in \partial (q_n^* h)(x_n), \\
\langle q_n^*, q_n'' - h(x_n) \rangle = 0 \; \forall n \in \mathbb{N}, u_n^* + u_n^{*\prime} \to 0, e_n^* \to 0, x_n \to a, p_n \to a, \\
y_n \to h(a), q_n \to h(a), q_n'' \to h(a) \; (n \to +\infty) \text{ and} \\
f(p_n) + g(q_n) - \langle u_n^*, p_n - x_n \rangle - \langle q_n^*, q_n - q_n'' \rangle - f(a) - g(h(a)) \to 0 \; (n \to +\infty).
\end{cases}
\tag{3. 21}
$$

Let us notice that the sequence $\{ y_n : n \in \mathbb{N} \}$ plays no role in (3. 21), which is thus equivalent to

$$
\begin{cases}
\exists (x_n, p_n, q_n, q_n'') \in X \times \mathrm{dom}\, f \times \mathrm{dom}\, g \times Y : h(x_n) \leq_C q_n'', \\
\exists (u_n^*, e_n^*, u_n^{*\prime}, q_n^*) : q_n^* \in C^*, u_n^* \in \partial f(p_n), q_n^* + e_n^* \in \partial g(q_n), u_n^{*\prime} \in \partial (q_n^* h)(x_n), \\
\langle q_n^*, q_n'' - h(x_n) \rangle = 0 \; \forall n \in \mathbb{N}, u_n^* + u_n^{*\prime} \to 0, e_n^* \to 0, x_n \to a, p_n \to a, \\
q_n \to h(a), q_n'' \to h(a) \; (n \to +\infty) \text{ and} \\
f(p_n) + g(q_n) - \langle u_n^*, p_n - x_n \rangle - \langle q_n^*, q_n - q_n'' \rangle - f(a) - g(h(a)) \to 0 \; (n \to +\infty).
\end{cases}
\tag{3. 22}
$$

Indeed, the direct implication is obvious, while for the reverse one we take $y_n := h(a)$ for all $n \in \mathbb{N}$.

Let us introduce now the following real sequences: $a_n := f(p_n) + g(q_n) - \langle u_n^*, p_n - x_n \rangle - \langle q_n^*, q_n - q_n'' \rangle - f(a) - g(h(a))$, $b_n := g(q_n) - \langle q_n^*, q_n - h(a) \rangle - g(h(a))$ and $c_n := f(p_n) - \langle u_n^*, p_n - x_n \rangle + \langle q_n^*, h(x_n) - h(a) \rangle - f(a)$ for all $n \in \mathbb{N}$. We prove that if the condition

$$
\begin{cases}
(x_n, p_n, q_n, q_n'') \in X \times \mathrm{dom}\, f \times \mathrm{dom}\, g \times Y, u_n^* \in \partial f(p_n), q_n^* + e_n^* \in \partial g(q_n), \\
u_n^{*\prime} \in \partial (q_n^* h)(x_n), \langle q_n^*, q_n'' - h(x_n) \rangle = 0 \; \forall n \in \mathbb{N} \text{ and} \\
u_n^* + u_n^{*\prime} \to 0, e_n^* \to 0, x_n \to a, q_n \to h(a) \; (n \to +\infty)
\end{cases}
\tag{3. 23}
$$

is satisfied, then we have

$$
a_n \to 0 \; (n \to +\infty) \text{ if and only if } b_n \to 0 \text{ and } c_n \to 0 \; (n \to +\infty).
\tag{3. 24}
$$

Indeed, if (3. 23) is fulfilled, then

$$
a_n = b_n + c_n,
\tag{3. 25}
$$

hence the sufficiency of relation (3. 24) is trivial. We point out that for this implication we need only the fulfillment of $\langle q_n^*, q_n'' - h(x_n) \rangle = 0$ for all $n \in \mathbb{N}$.

Assume now that $a_n \to 0$ $(n \to +\infty)$. Since $u_n^* \in \partial f(p_n)$ we have $f(a) - f(p_n) \geq \langle u_n^*, a - p_n \rangle$ for all $n \in \mathbb{N}$. Moreover, $u_n^{*\prime} \in \partial (q_n^* h)(x_n)$, hence $\langle q_n^*, h(a) \rangle - \langle q_n^*, h(x_n) \rangle \geq \langle u_n^{*\prime}, a - x_n \rangle$ for all $n \in \mathbb{N}$. We obtain that $c_n \leq \langle u_n^*, p_n - a \rangle + \langle u_n^{*\prime}, x_n - a \rangle - \langle u_n^*, p_n - x_n \rangle = \langle u_n^* + u_n^{*\prime}, x_n - a \rangle$. Also, from $q_n^* + e_n^* \in \partial g(q_n)$ we get $g(h(a)) - g(q_n) \geq \langle q_n^* + e_n^*, h(a) - q_n \rangle$ and so

$$
b_n \leq \langle q_n^* + e_n^*, q_n - h(a) \rangle - \langle q_n^*, q_n - h(a) \rangle = \langle e_n^*, q_n - h(a) \rangle.
$$

On the other hand, we have

$$
b_n = a_n - c_n \geq a_n - \langle u_n^* + u_n^{*\prime}, x_n - a \rangle.
$$

Combining the last two inequalities we obtain $b_n \to 0$ $(n \to +\infty)$. From (3. 25) we also get $c_n \to 0$ $(n \to +\infty)$ and hence (3. 24) is fulfilled.

Thus the condition (3. 22) is equivalent to (3. 17) and the proof is complete. $\square$

In the following corollary we give a sequential characterization of the subgradients of the function $g \circ h$ at $a \in \mathrm{dom}\, h \cap h^{-1}(\mathrm{dom}\, g)$.

Corollary 3.7 *For $a \in \operatorname{dom} h \cap h^{-1}(\operatorname{dom} g)$ we have $x^* \in \partial(g \circ h)(a)$ if and only if*

$$
\begin{cases}
\exists(x_n, q_n, q'_n) \in X \times \operatorname{dom} g \times Y, h(x_n) \leq_C q'_n, \exists(e_n^*, x_n^*, q_n^*), q_n^* \in C^*, \\
q_n^* + e_n^* \in \partial g(q_n), x_n^* \in \partial(q_n^* h)(x_n), \langle q_n^*, q'_n - h(x_n)\rangle = 0 \; \forall n \in \mathbb{N}, \\
x_n \to a, \; q_n \to h(a), \; q'_n \to h(a), \; x_n^* \to x^*, \; e_n^* \to 0 \; (n \to +\infty), \\
g(q_n) - \langle q_n^*, q_n - h(a)\rangle - g(h(a)) \to 0 \; (n \to +\infty) \text{ and} \\
\langle q_n^*, h(x_n) - h(a)\rangle \to 0 \; (n \to +\infty).
\end{cases}
\tag{3. 26}
$$

Proof. We have $x^* \in \partial(g \circ h)(a) \Leftrightarrow 0 \in \partial(-x^* + g \circ h)(a) \Leftrightarrow a$ is an optimal solution of the problem (P^{CC}) with $f : X \to \mathbb{R}$ defined by $f(x) = \langle -x^*, x \rangle$ for all $x \in X$. According to Theorem 3.8 we get that $x^* \in \partial(g \circ h)(a)$ if and only if

$$
\begin{cases}
\exists(x_n, p_n, q_n, q'_n) \in X \times X \times \operatorname{dom} g \times Y, h(x_n) \leq_C q'_n, \exists(e_n^*, u_n^{*\prime}, q_n^*), q_n^* \in C^*, \\
q_n^* + e_n^* \in \partial g(q_n), u_n^{*\prime} \in \partial(q_n^* h)(x_n), \langle q_n^*, q'_n - h(x_n)\rangle = 0 \; \forall n \in \mathbb{N}, \\
x_n \to a, \; p_n \to a, \; q_n \to h(a), \; q'_n \to h(a), \; u_n^{*\prime} \to x^*, \; e_n^* \to 0 \; (n \to +\infty), \\
g(q_n) - \langle q_n^*, q_n - h(a)\rangle - g(h(a)) \to 0 \; (n \to +\infty) \text{ and} \\
\langle q_n^*, h(x_n) - h(a)\rangle \to 0 \; (n \to +\infty),
\end{cases}
\tag{3. 27}
$$

where we used the continuity of the function f and the fact that $\partial f(x) = \{-x^*\}$ for all $x \in X$. The desired conclusion follows easily, since in the condition (3. 27) the sequence $\{p_n : n \in \mathbb{N}\}$ is superfluous (we made the notation $x_n^* := u_n^{*\prime}$ for all $n \in \mathbb{N}$). $\qquad \square$

Remark 3.9 Corollary 3.7 above is exactly the result given by THIBAULT in [138, Theorem 3.1].

3.4.2   The case h is continuous

We consider again the setting from the beginning of Section 3.4, with the additional hypotheses that $h : X \to Y$ is continuous and $g : Y \to \overline{\mathbb{R}}$ is C-increasing on Y. We want to mention that, unlike in the previous subsection, the results in this subsection hold even in the case Y is not reflexive. We consider the perturbation function $\Phi^{CC} : X \times Y \to \overline{\mathbb{R}}$,

$$
\Phi^{CC}(x, y) = f(x) + g(h(x) + y) \; \forall(x, y) \in X \times Y, \tag{3. 28}
$$

which is in this situation proper, convex and lower semicontinuous. The conjugate function $(\Phi^{CC})^* : X^* \times Y^* \to \overline{\mathbb{R}}$ has for all $(x^*, y^*) \in X^* \times Y^*$ the following form

$$
(\Phi^{CC})^*(x^*, y^*) = \begin{cases} (f + y^* h)^*(x^*) + g^*(y^*), & \text{if } y^* \in C^*, \\ +\infty, & \text{otherwise,} \end{cases}
$$

where we took into consideration that $g^*(y^*) = +\infty$ for all $y^* \in Y^* \setminus C^*$. By means of the general result Theorem 3.2 applied for this perturbation function we obtain the following sequential optimality conditions for (P^{CC}).

Theorem 3.9 *The element $a \in \operatorname{dom} f \cap h^{-1}(\operatorname{dom} g)$ is an optimal solution of the problem (P^{CC}) if and only if*

$$
\begin{cases}
\exists(x_n, y_n) \in \operatorname{dom} f \times \operatorname{dom} g, \exists(u_n^*, v_n^*, y_n^*) \in X^* \times X^* \times C^*, \\
u_n^* \in \partial f(x_n), v_n^* \in \partial(y_n^* h)(x_n), y_n^* \in \partial g(y_n) \; \forall n \in \mathbb{N}, \\
u_n^* + v_n^* \to 0, \; x_n \to a, \; y_n \to h(a) \; (n \to +\infty), \\
f(x_n) + \langle y_n^*, h(x_n) - h(a)\rangle - f(a) \to 0 \; (n \to +\infty) \text{ and} \\
g(y_n) - \langle y_n^*, y_n - h(a)\rangle - g(h(a)) \to 0 \; (n \to +\infty).
\end{cases}
\tag{3. 29}
$$

Proof. Applying Theorem 3.2 we obtain that $a \in \operatorname{dom} f \cap h^{-1}(\operatorname{dom} g)$ is an optimal solution of the problem (P^{CC}) if and only if

$$\begin{cases} \exists (x_n, y_n) \in X \times Y, x_n \in \operatorname{dom} f, h(x_n) + y_n \in \operatorname{dom} g, \\ \exists (x_n^*, y_n^*) \in \partial \Phi^{CC}(x_n, y_n) \ \forall n \in \mathbb{N}, x_n^* \to 0, x_n \to a, y_n \to 0 \ (n \to +\infty) \text{ and} \\ \Phi^{CC}(x_n, y_n) - \langle y_n^*, y_n \rangle - \Phi^{CC}(a, 0) \to 0 \ (n \to +\infty). \end{cases}$$
(3. 30)

The condition $(x_n^*, y_n^*) \in \partial \Phi^{CC}(x_n, y_n)$ is equivalent to $y_n^* \in C^*$ and $f(x_n) + g(h(x_n) + y_n) + (f + y_n^* h)^*(x_n^*) + g^*(y_n^*) = \langle x_n^*, x_n \rangle + \langle y_n^*, y_n \rangle$ for all $n \in \mathbb{N}$. By using the Young-Fenchel inequality one can see that for all $n \in \mathbb{N}$ we have

$$f(x_n) + (y_n^* h)(x_n) + (f + y_n^* h)^*(x_n^*) - \langle x_n^*, x_n \rangle \geq 0$$

and

$$g(h(x_n) + y_n) + g^*(y_n^*) - \langle y_n^*, h(x_n) + y_n \rangle \geq 0.$$

Since the sum of the terms in the left-hand side of the inequalities above is equal to zero, both of them must be equal to zero. This is the case if and only if $x_n^* \in \partial(f + y_n^* h)(x_n)$ and $y_n^* \in \partial g(h(x_n) + y_n)$ for all $n \in \mathbb{N}$. Hence $a \in \operatorname{dom} f \cap h^{-1}(\operatorname{dom} g)$ is an optimal solution of (P^{CC}) if and only if

$$\begin{cases} \exists (x_n, y_n) \in X \times Y, x_n \in \operatorname{dom} f, h(x_n) + y_n \in \operatorname{dom} g, \\ \exists (x_n^*, y_n^*) \in X^* \times C^*, x_n^* \in \partial(f + y_n^* h)(x_n), y_n^* \in \partial g(h(x_n) + y_n) \ \forall n \in \mathbb{N}, \\ x_n^* \to 0, \ x_n \to a, \ y_n \to 0 \ (n \to +\infty) \text{ and} \\ f(x_n) + g(h(x_n) + y_n) - \langle y_n^*, y_n \rangle - f(a) - g(h(a)) \to 0 \ (n \to +\infty). \end{cases}$$
(3. 31)

The function h being continuous, the following subdifferential sum formula holds:

$$\partial(f + y_n^* h)(x_n) = \partial f(x_n) + \partial(y_n^* h)(x_n) \ \forall n \in \mathbb{N} \tag{3. 32}$$

(cf. [149, Theorem 2.8.7]). Thus $x_n^* \in \partial(f + y_n^* h)(x_n)$ if and only if there exist $u_n^* \in \partial f(x_n)$ and $v_n^* \in \partial(y_n^* h)(x_n)$ such that $x_n^* = u_n^* + v_n^*$ for all $n \in \mathbb{N}$. Introducing a new variable by $y_n' := h(x_n) + y_n$ for all $n \in \mathbb{N}$ and employing once more the continuity of the function h we get that (3. 31) is equivalent to

$$\begin{cases} \exists (x_n, y_n') \in \operatorname{dom} f \times \operatorname{dom} g, \exists (u_n^*, v_n^*, y_n^*) \in X^* \times X^* \times C^*, \\ u_n^* \in \partial f(x_n), v_n^* \in \partial(y_n^* h)(x_n), y_n^* \in \partial g(y_n') \ \forall n \in \mathbb{N}, \\ u_n^* + v_n^* \to 0, \ x_n \to a, \ y_n' \to h(a) \ (n \to +\infty) \text{ and} \\ f(x_n) + g(y_n') - \langle y_n^*, y_n' - h(x_n) \rangle - f(a) - g(h(a)) \to 0 \ (n \to +\infty). \end{cases}$$
(3. 33)

Let us consider now the following real sequences: $\alpha_n := f(x_n) + g(y_n') - \langle y_n^*, y_n' - h(x_n) \rangle - f(a) - g(h(a))$, $\beta_n := f(x_n) - f(a) + \langle y_n^*, h(x_n) - h(a) \rangle$ and $\gamma_n := g(y_n') - g(h(a)) - \langle y_n^*, y_n' - h(a) \rangle$ for all $n \in \mathbb{N}$. We have $\alpha_n = \beta_n + \gamma_n$ for all $n \in \mathbb{N}$ and if the condition

$$\begin{cases} (x_n, y_n') \in \operatorname{dom} f \times \operatorname{dom} g, (u_n^*, v_n^*, y_n^*) \in X^* \times X^* \times C^*, \\ u_n^* \in \partial f(x_n), v_n^* \in \partial(y_n^* h)(x_n), y_n^* \in \partial g(y_n') \ \forall n \in \mathbb{N}, \\ u_n^* + v_n^* \to 0, \ x_n \to a, \ (n \to +\infty), \end{cases}$$
(3. 34)

is satisfied, then

$$\alpha_n \to 0 \ (n \to +\infty) \text{ if and only if } \beta_n \to 0 \text{ and } \gamma_n \to 0 \ (n \to +\infty). \tag{3. 35}$$

We omit the proof of (3. 35), since it can be done in the lines of the one given for the relation (3. 24) in the proof of Theorem 3.8. Hence the condition (3. 33) is equivalent to (3. 29). □

Remark 3.10 Particularizing Theorem 3.9 to the case $C = \{0\}$ and $h = A$, where $A : X \to Y$ is a continuous linear mapping, we rediscover Theorem 3.4. The details are left to the reader.

For the special case when $f = -x^*$, where $x^* \in X^*$ is fixed, we obtain from Theorem 3.9 the following result concerning the subgradients of the function $g \circ h$.

Corollary 3.8 *For* $a \in h^{-1}(\mathrm{dom}\, g)$ *we have* $x^* \in \partial(g \circ h)(a)$ *if and only if*

$$\begin{cases} \exists (x_n, y_n) \in X \times \mathrm{dom}\, g, \exists (v_n^*, y_n^*) \in X^* \times C^*, v_n^* \in \partial(y_n^* h)(x_n), y_n^* \in \partial g(y_n), \\ v_n^* \to x^*, \ x_n \to a, \ y_n \to h(a) \ (n \to +\infty), \\ g(y_n) - \langle y_n^*, y_n - h(a) \rangle - g(h(a)) \to 0 \ (n \to +\infty) \ and \\ \langle y_n^*, h(x_n) - h(a) \rangle \to 0 \ (n \to +\infty). \end{cases}$$

$$(3.\ 36)$$

Remark 3.11 The above sequential characterization of an arbitrary $x^* \in \partial(g \circ h)(a)$ was given by THIBAULT in case X and Y are reflexive Banach spaces, C is a closed, convex and normal cone and g is C-increasing on $h(X) + C$ (see [138, Corollary 3.2]). We proved that if the function g is C-increasing on Y, then this result holds even if the cone C is not normal and Y is an arbitrary Banach space. Moreover, the closedness condition regarding the cone C, requested by THIBAULT in [138, Corollary 3.2], is not needed anymore in this situation.

Remark 3.12 (a) One can prove that the perturbation function defined at the beginning of subsection 3.4.2 is lower semicontinuous even in the more general case when h is star C-lower semicontinuous (this is a direct consequence of the equality $(\Phi^{CC})^{**}(x, y) = \Phi^{CC}(x, y)$ for all $(x, y) \in X \times Y$, which can be proved by direct calculations, see also [35, Section 3]). This means that it is possible to derive sequential optimality conditions even in this case. Nevertheless, in order to obtain the result given by THIBAULT in [138, Corollary 3.2], we have to suppose that h is continuous, as this fact was used twice in the proof of Theorem 3.9 above. Even if the subdifferential sum formula (3. 32) holds also in the case h is star C-lower semicontinuous and f is continuous (because we take $f = -x^*$ in order to obtain the result of THIBAULT), we still need the continuity of the function h in order to ensure that the sequence y_n' has the limit $h(a)$ as $n \to +\infty$ (see the equivalence between the conditions (3. 31) and (3. 33) in the proof of Theorem 3.9).

(b) Under the hypotheses mentioned at the beginning of subsection 3.4.1 one cannot prove that the perturbation function Φ^{CC} defined in the relation (3. 28) is lower semicontinuous and hence in case h is C-epi-closed, Theorem 3.2 is not applicable for this perturbation function. This is one of the reasons why the first sequential optimality condition for the composed convex optimization problem (P^{CC}), namely Theorem 3.8, is derived via Corollary 3.6, a result which is given for an optimization problem with cone constraints (of course, Corollary 3.6 is obtained from the general result Theorem 3.2). Another reason is that the condition g is C-increasing on $h(\mathrm{dom}(h)) + C$ (which is the case in subsection 3.4.1) is not sufficient in order to guarantee the convexity of the above mentioned perturbation function. In order to ensure the convexity of the function Φ^{CC}, g has to be C-increasing on Y, which is actually the case in subsection 3.4.2.

Remark 3.13 Let us notice that by using the general sequential optimality conditions given in Remark 3.3, several sequential characterizations of optimal solutions for composed optimization problems with geometric and cone constraints are obtained in [31]. These conditions are then applied to equivalently characterize the properly efficient solutions (in the sense of linear scalarization) of vector optimization problems with geometric and cone constraints.

Chapter 4

Applications of the duality theory to enlargements of maximal monotone operators

Due to its applications in the theory of partial differential equations, the maximal monotonicity of operators defined on Banach spaces has been intensively studied since the beginning of this theory in the 1960's. We mention here the papers of BROWDER [42], MINTY [102] and ROCKAFELLAR [124, 125] who made the first important steps in this field. A comprehensive study of the theory of monotone operators can be found in the monographs of SIMONS [128, 130] and the lecture notes due to PHELPS [114], which are relevant references for anyone interested in this topic.

To an arbitrary monotone operator defined on a Banach space, FITZPATRICK associated in 1988 a convex and lower semicontinuous function which, in case the operator is maximally monotone, provides a characterization of the graph of the operator (cf. [65]). In this way, the theory of maximal monotone operators is linked with convex analysis. Unfortunately, these properties were not exploited until 2001, when the *Fitzpatrick function* was rediscovered by MARTÍNEZ-LEGAZ AND THÉRA in [100], and independently, by BURACHIK AND SVAITER in [51]. Since then, convex analysis, and in particular conjugate duality, plays an important role in the theory of maximal monotone operators.

Enlargements of maximal monotone operators were introduced and studied in order to find the zeros of a maximal monotone operator, that is to solve the problem

$$\text{find } x \in X \text{ such that } 0 \in S(x), \qquad (4.\ 1)$$

where $S : X \rightrightarrows X^*$ is a monotone operator and X is a Banach space. It is useful in this sense to consider an "enlargement" of S, that is a set-valued operator $S' : \mathbb{R}_+ \times X \rightrightarrows X^*$ with the property that $S(x) \subseteq S'(\varepsilon, x)$ for all $\varepsilon \geq 0$ and $x \in X$. Several examples of enlargements where introduced in the literature with the help of which one can develop algorithms to solve the problem (4. 1) (the first one was introduced by BURACHIK, IUSEM AND SVAITER in [46] with applications to variational inequalities). We refer to the monograph of BURACHIK AND IUSEM [45] for more details regarding enlargements of monotone operators and their applications.

We give in this chapter applications of the duality theory to enlargements of maximal monotone operators, proving once more the usefulness of convex analysis in the theory of monotone operators. Motivated by a classical result concerning the ε-subdifferential of the sum of two proper, convex and lower semicontinuous functions, we generalize this result to the context of enlargements of maximal monotone operators. An answer to an open problem posed by BURACHIK AND IUSEM is given and finally we introduce a regularity condition which ensures that the sum of the images of the enlargements of two maximal monotone operators is weak*-closed. The main results of this chapter are Corollary 4.2 (which is then applied in Section 4.3), Theorem 4.5, Corollary 4.6 and Theorem 4.8. The results presented in this chapter are based on [22, 24].

4.1 A bivariate infimal convolution formula

For the problem of establishing the maximal monotonicity of the sum of two maximal monotone operators defined on a reflexive Banach space it is important to give sufficient conditions for a so-called *bivariate infimal convolution formula* (see [113, 132]). This approach is useful when one tries to give a formula for the ε-enlargement of the sum of two maximal monotone operators (see Section 4.3). This is the reason why in the following we focus our attention on conditions which ensure this bivariate infimal convolution formula (see Corollary 4.2).

Let us consider X and Y separated locally convex spaces and X^*, Y^* their topological dual spaces, respectively. Part (i) of the next lemma plays an important role in deriving an equivalent characterization of the bivariate infimal convolution formula. Part (ii) finds applications in Section 4.5.

Lemma 4.1 *Let* $\Phi : X \times Y \to \overline{\mathbb{R}}$ *be a proper, convex and lower semicontinuous function.*

(i) If $0 \in \mathrm{pr}_Y(\mathrm{dom}\,\Phi)$ *then*

$$(\Phi(\cdot,0))^* = \mathrm{cl}_{w^*}\left(\inf_{y^* \in Y^*} \Phi^*(\cdot,y^*)\right) \tag{4. 2}$$

and

$$\mathrm{epi}(\Phi(\cdot,0))^* = \mathrm{cl}_{w^* \times \mathcal{R}}\left(\mathrm{pr}_{X^* \times \mathbb{R}}(\mathrm{epi}\,\Phi^*)\right). \tag{4. 3}$$

(ii) For all $x \in \mathrm{pr}_X(\mathrm{dom}\,\Phi)$ *we have*

$$\mathrm{pr}_{Y^*}(\mathrm{dom}\,\Phi^*) \subseteq \mathrm{dom}(\Phi(x,\cdot))^* \subseteq \mathrm{cl}_{w^*}\left(\mathrm{pr}_{Y^*}(\mathrm{dom}\,\Phi^*)\right). \tag{4. 4}$$

Proof. For (i) we refer to [35, Theorem 1 and Theorem 2]. Let us notice that relation (4. 2) was observed also in [150, page 197] and [112, pp. 628–629].

(ii) Let $x \in \mathrm{pr}_X(\mathrm{dom}\,\Phi)$ be fixed and define $\Psi : Y \times X \to \overline{\mathbb{R}}$ by $\Psi(y,u) = \Phi(x+u,y)$ for all $(y,u) \in Y \times X$. The function Ψ is proper, convex and lower semicontinuous and fulfills $\Psi(y,0) = \Phi(x,y)$ for all $y \in Y$. Since $x \in \mathrm{pr}_X(\mathrm{dom}\,\Phi)$ one has that $0 \in \mathrm{pr}_X(\mathrm{dom}\,\Psi)$.

According to (4. 2) we have that

$$(\Psi(\cdot,0))^* = \mathrm{cl}_{w^*}\left(\inf_{x^* \in X^*} \Psi^*(\cdot,x^*)\right).$$

Consequently,

$$\mathrm{dom}\left(\inf_{x^* \in X^*} \Psi^*(\cdot,x^*)\right) \subseteq \mathrm{dom}\left(\mathrm{cl}_{w^*}\left(\inf_{x^* \in X^*} \Psi^*(\cdot,x^*)\right)\right)$$

$$\subseteq \mathrm{cl}_{w^*}\left(\mathrm{dom}\left(\inf_{x^* \in X^*} \Psi^*(\cdot,x^*)\right)\right).$$

Further we have that $v^* \in \mathrm{dom}\,(\inf_{x^* \in X^*} \Psi^*(\cdot,x^*))$ if and only if there exists $x^* \in X^*$ such that $\Psi^*(v^*,x^*) < +\infty$. Since

$$\Psi^*(v^*,x^*) = \sup_{v \in Y, u \in X}\{\langle v^*,v\rangle + \langle x^*,u\rangle - \Phi(x+u,v)\}$$

$$= \sup_{v \in Y, t \in X} \{\langle v^*, v \rangle + \langle x^*, t - x \rangle - \Phi(t, v)\} = -\langle x^*, x \rangle + \Phi^*(x^*, v^*),$$

this is the same with having that $\Phi^*(x^*, v^*) < +\infty$ or, equivalently, $(x^*, v^*) \in$ dom Φ^*. Therefore dom $(\inf_{x^* \in X^*} \Psi^*(\cdot, x^*)) = \operatorname{pr}_{Y^*}(\operatorname{dom} \Phi^*)$ and the conclusion follows. □

Before we proceed, let us recall the following concept. For M, Z two subsets of X, we say that M is *closed regarding* the set Z if $M \cap Z = \operatorname{cl}(M) \cap Z$. It is worth noting that a closed set is closed regarding any set. Several weak regularity conditions (in the theory of maximal monotone operators and convex optimization) are expressed by using this notion, see [27, 32, 33, 35].

Theorem 4.1 *Let* $\Phi : X \times Y \to \overline{\mathbb{R}}$ *be a proper, convex and lower semicontinuous function such that* $0 \in \operatorname{pr}_Y(\operatorname{dom} \Phi)$ *and* U *be a non-empty subset of* X^*. *Then the following statements are equivalent:*

(i) $\sup_{x \in X}\{\langle x^*, x \rangle - \Phi(x, 0)\} = \min_{y^* \in Y^*} \Phi^*(x^*, y^*)$ *for all* $x^* \in U$;

(ii) $\operatorname{pr}_{X^* \times \mathbb{R}}(\operatorname{epi} \Phi^*)$ *is closed regarding* $U \times \mathbb{R}$ *in* $(X^*, \omega(X^*, X)) \times \mathbb{R}$.

Proof. (i)$\Rightarrow$(ii) Take an arbitrary element $(x^*, r) \in \operatorname{cl}_{w^* \times \mathcal{R}} \left(\operatorname{pr}_{X^* \times \mathbb{R}}(\operatorname{epi} \Phi^*) \right) \cap$ $(U \times \mathbb{R})$. Lemma 4.1(i) guarantees that $(x^*, r) \in \operatorname{epi}(\Phi(\cdot, 0))^*$, which implies $(\Phi(\cdot, 0))^*(x^*) \leq r$, that is $\sup_{x \in X}\{\langle x^*, x \rangle - \Phi(x, 0)\} \leq r$. From (i) we obtain the existence of an element $y^* \in Y^*$ such that $\Phi^*(x^*, y^*) \leq r$, thus $(x^*, r) \in$ $\operatorname{pr}_{X^* \times \mathbb{R}}(\operatorname{epi} \Phi^*) \cap (U \times \mathbb{R})$. Hence we have $\operatorname{cl}_{w^* \times \mathcal{R}} \left(\operatorname{pr}_{X^* \times \mathbb{R}}(\operatorname{epi} \Phi^*) \right) \cap (U \times \mathbb{R}) \subseteq$ $\operatorname{pr}_{X^* \times \mathbb{R}}(\operatorname{epi} \Phi^*) \cap (U \times \mathbb{R})$, and since the reverse inclusion is always satisfied, we obtain that (ii) is fulfilled.

Conversely, suppose that (ii) is true and take $x^* \in U$ arbitrary. From the Young-Fenchel inequality we obtain

$$(\Phi(\cdot, 0))^*(x^*) \leq \inf_{y^* \in Y^*} \Phi^*(x^*, y^*). \tag{4.5}$$

In case $(\Phi(\cdot, 0))^*(x^*) = +\infty$, then (i) is obviously satisfied. So we may suppose that $(\Phi(\cdot, 0))^*(x^*) < +\infty$. Taking into consideration that $0 \in \operatorname{pr}_Y(\operatorname{dom} \Phi)$ we easily derive that $(\Phi(\cdot, 0))^*(x^*) \in \mathbb{R}$. We get by relation (4.3) and (ii) that $(x^*, (\Phi(\cdot, 0))^*(x^*)) \in \operatorname{epi}(\Phi(\cdot, 0))^* \cap (U \times \mathbb{R}) = \operatorname{cl}_{w^* \times \mathcal{R}} \left(\operatorname{pr}_{X^* \times \mathbb{R}}(\operatorname{epi} \Phi^*) \right) \cap (U \times \mathbb{R}) =$ $\left(\operatorname{pr}_{X^* \times \mathbb{R}}(\operatorname{epi} \Phi^*) \right) \cap (U \times \mathbb{R})$. Hence there exists an element $\overline{y}^* \in Y^*$ such that $\Phi^*(x^*, \overline{y}^*) \leq (\Phi(\cdot, 0))^*(x^*)$. Combining this with (4.5) we obtain $(\Phi(\cdot, 0))^*(x^*) =$ $\Phi^*(x^*, \overline{y}^*) = \min_{y^* \in Y^*} \Phi^*(x^*, y^*)$. As $x^* \in U$ was arbitrary taken, the proof is complete. □

Remark 4.1 One can give an alternative proof of the above theorem. To this end, let (X, τ) be a topological space, where τ is the corresponding topology on X, $U \subseteq X$ and $A \subseteq X \times \mathbb{R}$. One can prove that $A \cap (U \times \mathbb{R}) = \operatorname{cl}_{\tau \times \mathcal{R}}(A) \cap$ $(U \times \mathbb{R})$ if and only if $A \cap (\{u\} \times \mathbb{R}) = \operatorname{cl}_{\tau \times \mathcal{R}}(A) \cap (\{u\} \times \mathbb{R})$ for all $u \in U$. Using this remark, one can deduce Theorem 4.1 from the corresponding statement with U a singleton. Indeed, for $U = \{x^*\}$, the statement (i) is nothing else than $(\Phi(\cdot, 0))^*(x^*) = \operatorname{cl}_{w^*} \left(\inf_{y^* \in Y^*} \Phi^*(\cdot, y^*) \right)(x^*) = \inf_{y^* \in Y^*} \Phi^*(x^*, y^*)$ and the infimum is attained, while (ii) asserts that for $A := \operatorname{pr}_{X^* \times \mathbb{R}}(\operatorname{epi} \Phi^*)$ one has $A \cap (\{x^*\} \times \mathbb{R}) = \operatorname{cl}_{w^* \times \mathcal{R}}(A) \cap (\{x^*\} \times \mathbb{R})$. For $x^* = 0$ the equivalence of (i) and (ii) is nothing else than [116, Theorem 4.3.1] (see also [115, pp. 6]). The statement for $x^* \neq 0$ can be deduced by making a translation.

Remark 4.2 Considering in the previous theorem $U := X^*$ we obtain, under the same hypotheses as in Theorem 4.1, that the following conditions are equivalent:

(i) $\sup_{x \in X} \{ \langle x^*, x \rangle - \Phi(x, 0) \} = \min_{y^* \in Y^*} \Phi^*(x^*, y^*)$ for all $x^* \in X^*$;

(ii) $\mathrm{pr}_{X^* \times \mathbb{R}}(\mathrm{epi}\,\Phi^*)$ is closed in $(X^*, \omega(X^*, X)) \times \mathbb{R}$.

This statement can be deduced from [117, Theorem 2.2] and it was proved in [48] in the case of Banach spaces and in [35] in the framework of separated locally convex spaces. Let us notice that in the literature condition (i) is referred to as *stable strong duality* (see [21, 35, 48, 130] for more details).

An important special case of Theorem 4.1 follows.

Corollary 4.1 *Let $f, g : X \to \overline{\mathbb{R}}$ be proper, convex and lower semicontinuous functions such that $\mathrm{dom}\,f \cap \mathrm{dom}\,g \neq \emptyset$ and U be a non-empty subset of X^*. Then the following statements are equivalent:*

(i) $(f + g)^*(x^*) = (f^* \square g^*)(x^*)$ and $f^* \square g^*$ is exact at x^* for all $x^* \in U$;

(ii) $\mathrm{epi}\,f^* + \mathrm{epi}\,g^*$ is closed regarding $U \times \mathbb{R}$ in $(X^*, \omega(X^*, X)) \times \mathbb{R}$.

Proof. Consider the function $\Phi : X \times X \to \overline{\mathbb{R}}$ defined by $\Phi(x, y) = f(x) + g(x + y)$ for all $(x, y) \in X \times X$. A simple computation shows that $\Phi^*(x^*, y^*) = f^*(x^* - y^*) + g^*(y^*)$ for all $(x^*, y^*) \in X^* \times X^*$. One can prove easily that the hypotheses of Theorem 4.1 are satisfied for this particular choice of the function Φ. The result follows now by applying Theorem 4.1.   □

Remark 4.3 In case $U = X^*$, the previous corollary was established by BURACHIK AND JEYAKUMAR in Banach spaces (cf. [47, Theorem 1]) and by BOŢ AND WANKA in separated locally convex spaces (cf. [39, Theorem 3.2]).

The following result will lead us to the *bivariate inf-convolution formula*.

Theorem 4.2 *Let $h_1, h_2 : X \times Y \to \overline{\mathbb{R}}$ be proper, convex and lower semicontinuous functions such that $\mathrm{pr}_X(\mathrm{dom}\,h_1) \cap \mathrm{pr}_X(\mathrm{dom}\,h_2) \neq \emptyset$ and V be a non-empty subset of Y^*. Consider the functions $h_1 \square_2 h_2 : X \times Y \to \overline{\mathbb{R}}$, $(h_1 \square_2 h_2)(x, y) = \inf\{h_1(x, u) + h_2(x, v) : u, v \in Y, u + v = y\}$ and $h_1^* \square_1 h_2^* : X^* \times Y^* \to \overline{\mathbb{R}}$, $(h_1^* \square_1 h_2^*)(x^*, y^*) = \inf\{h_1^*(u^*, y^*) + h_2^*(v^*, y^*) : u^*, v^* \in X^*, u^* + v^* = x^*\}$. Then the following conditions are equivalent:*

(i) $(h_1 \square_2 h_2)^*(x^*, y^*) = (h_1^* \square_1 h_2^*)(x^*, y^*)$ and $h_1^* \square_1 h_2^*$ is exact at (x^*, y^*) (that is, the infimum in the definition of $(h_1^* \square_1 h_2^*)(x^*, y^*)$ is attained) for all $(x^*, y^*) \in X^* \times V$;

(ii) $\{(a^* + b^*, u^*, v^*, r) : h_1^*(a^*, u^*) + h_2^*(b^*, v^*) \leq r\}$ is closed regarding the set $X^* \times \Delta_V \times \mathbb{R}$ in $(X^*, \omega(X^*, X)) \times (Y^*, \omega(Y^*, Y)) \times (Y^*, \omega(Y^*, Y)) \times \mathbb{R}$, where $\Delta_V = \{(y^*, y^*) : y^* \in V\}$.

Proof. Take an arbitrary $(x^*, y^*) \in X^* \times Y^*$. The following equality can be easily derived

$$(h_1 \square_2 h_2)^*(x^*, y^*) = \sup_{x \in X, u, v \in Y} \{ \langle x^*, x \rangle + \langle y^*, u + v \rangle - h_1(x, u) - h_2(x, v) \}. \quad (4.\,6)$$

Define now the functions $F, G : X \times Y \times Y \to \overline{\mathbb{R}}$, by $F(x, u, v) = h_1(x, u)$ and $G(x, u, v) = h_2(x, v)$ for all $(x, u, v) \in X \times Y \times Y$. It holds $(h_1 \square_2 h_2)^*(x^*, y^*) = (F + G)^*(x^*, y^*, y^*)$. One can show that for all $(x^*, u^*, v^*) \in X^* \times Y^* \times Y^*$, the conjugate functions $F^*, G^* : X^* \times Y^* \times Y^* \to \overline{\mathbb{R}}$ have the following formulations

$$F^*(x^*, u^*, v^*) = \begin{cases} h_1^*(x^*, u^*), & \text{if } v^* = 0, \\ +\infty, & \text{otherwise} \end{cases}$$

and

$$G^*(x^*, u^*, v^*) = \begin{cases} h_2^*(x^*, v^*), & \text{if } u^* = 0, \\ +\infty, & \text{otherwise,} \end{cases}$$

respectively. Further we have $(F^* \Box G^*)(x^*, y^*, y^*) = (h_1^* \Box_1 h_2^*)(x^*, y^*)$. Hence the condition (i) is fulfilled if and only if $(F+G)^*(x^*, y^*, y^*) = (F^* \Box G^*)(x^*, y^*, y^*)$ and $(F^* \Box G^*)(x^*, y^*, y^*)$ is exact at (x^*, y^*, y^*) for all $(x^*, y^*, y^*) \in X^* \times \Delta_V$. In view of Corollary 4.1, the last condition is equivalent to epi F^* + epi G^* is closed regarding the set $X^* \times \Delta_V \times \mathbb{R}$ in $(X^*, \omega(X^*, X)) \times (Y^*, \omega(Y^*, Y)) \times (Y^*, \omega(Y^*, Y)) \times \mathbb{R}$. Finally the equality epi F^* + epi $G^* = \{(a^* + b^*, u^*, v^*, r) : h_1^*(a^*, u^*) + h_2^*(b^*, v^*) \leq r\}$, whose proof presents no difficulty, gives the desired result. □

For the particular case when $V := Y^*$ we obtain a necessary and sufficient condition for the bivariate infimal convolution formula (relation (i) in the result below).

Corollary 4.2 *Let $h_1, h_2 : X \times Y \to \overline{\mathbb{R}}$ be proper, convex and lower semicontinuous functions such that $\mathrm{pr}_X(\mathrm{dom}\, h_1) \cap \mathrm{pr}_X(\mathrm{dom}\, h_2) \neq \emptyset$. The following statements are equivalent:*

(i) $(h_1 \Box_2 h_2)^* = h_1^* \Box_1 h_2^*$ and $h_1^* \Box_1 h_2^*$ is exact;

(ii) $\{(a^* + b^*, u^*, v^*, r) : h_1^*(a^*, u^*) + h_2^*(b^*, v^*) \leq r\}$ is closed regarding the set $X^* \times \Delta_{Y^*} \times \mathbb{R}$ in $(X^*, \omega(X^*, X)) \times (Y^*, \omega(Y^*, Y)) \times (Y^*, \omega(Y^*, Y)) \times \mathbb{R}$.

Remark 4.4 A generalized interior-point condition which guarantees relation (i) in Corollary 4.2 (and implicitly also (ii)) has been given by SIMONS AND ZĂLINESCU in the framework of Banach spaces (cf. [132, Theorem 4.2]), namely:

$$(CQ^{SZ}) \qquad 0 \in \mathrm{sqri}\left(\mathrm{pr}_X(\mathrm{dom}\, h_1) - \mathrm{pr}_X(\mathrm{dom}\, h_2)\right).$$

Nevertheless, unlike the condition (ii), which is necessary and sufficient for (i), the condition (CQ^{SZ}) is only sufficient, as the following example, which can be found in [27], shows.

Example 4.1 Take $X = Y = \mathbb{R}^2$, equipped with the Euclidean norm $\|\cdot\|_2$, $f, g : \mathbb{R}^2 \to \overline{\mathbb{R}}$, $f = \|\cdot\|_2 + \delta_{\mathbb{R}^2_+}$, $g = \delta_{-\mathbb{R}^2_+}$,

$$h_1(x, x^*) = f(x) + f^*(x^*) \text{ for all } (x, x^*) \in \mathbb{R}^2 \times \mathbb{R}^2$$

and, respectively,

$$h_2(x, x^*) = g(x) + g^*(x^*) \text{ for all } (x, x^*) \in \mathbb{R}^2 \times \mathbb{R}^2.$$

One can see that $g^* = \delta_{\mathbb{R}^2_+}$ and $f^* = \delta_{\overline{B}(0,1) - \mathbb{R}^2_+}$, where $\overline{B}(0, 1)$ is the closed unit ball of $\mathbb{R}^2$. We have

$$\{(x^* + y^*, x, y, r) : f(x) + f^*(x^*) + g(y) + g^*(y^*) \leq r\} =$$

$$\mathbb{R}^2 \times \{(x, y, r) : x \in \mathbb{R}^2_+, y \in -\mathbb{R}^2_+, \|x\|_2 \leq r\},$$

which is closed, hence closed regarding the set $\mathbb{R}^2 \times \Delta_{\mathbb{R}^2} \times \mathbb{R}$. Thus, by Corollary 4.2, (i) is fulfilled. However, condition (CQ^{SZ}) becomes: $\mathbb{R}^2_+$ is a closed linear subspace of $\mathbb{R}^2$, which is of course a false statement.

By taking in Theorem 4.2 $Y = X^*$ and $V = X$, where X is supposed to be a normed space (in this case $V = X$ can be seen as a subspace of $Y^* = X^{**}$), we obtain the following result.

Corollary 4.3 *Let $h_1, h_2 : X \times X^* \to \overline{\mathbb{R}}$ be proper, convex and lower semicontinuous functions in the strong topology of $X \times X^*$ such that $\mathrm{pr}_X(\mathrm{dom}\, h_1) \cap \mathrm{pr}_X(\mathrm{dom}\, h_2) \neq \emptyset$. The following statements are equivalent:*

(i) *$(h_1 \square_2 h_2)^*(x^*, x) = (h_1^* \square_1 h_2^*)(x^*, x)$ and $h_1^* \square_1 h_2^*$ is exact at (x^*, x) for all $(x^*, x) \in X^* \times X$;*

(ii) *$\{(a^* + b^*, u^{**}, v^{**}, r) : h_1^*(a^*, u^{**}) + h_2^*(b^*, v^{**}) \leq r\}$ is closed regarding the set $X^* \times \Delta_X \times \mathbb{R}$ in $(X^*, \omega(X^*, X)) \times (X^{**}, \omega(X^{**}, X^*)) \times (X^{**}, \omega(X^{**}, X^*)) \times \mathbb{R}$.*

4.2   Monotone operators and enlargements

In this section we recall some notations and results concerning monotone operators and enlargements. Consider further X a nontrivial Banach space, X^* its topological dual space and X^{**} its bidual space. A set-valued operator $S : X \rightrightarrows X^*$ is said to be *monotone* if

$$\langle y^* - x^*, y - x \rangle \geq 0, \ \text{whenever} \ y^* \in S(y) \ \text{and} \ x^* \in S(x).$$

The monotone operator S is called *maximal monotone* if its graph

$$G(S) = \{(x, x^*) : x^* \in S(x)\} \subseteq X \times X^*$$

is not properly contained in the graph of any other monotone operator $S' : X \rightrightarrows X^*$. For S we consider also its *domain* $D(S) = \{x \in X : S(x) \neq \emptyset\} = \mathrm{pr}_X(G(S))$ and its *range* $R(S) = \cup_{x \in X} S(x) = \mathrm{pr}_{X^*}(G(S))$. The classical example of a maximal monotone operator is the subdifferential of a proper, convex and lower semicontinuous function (this result is due to Rockafellar, cf. [124]). However, there exist maximal monotone operators which are not subdifferentials (cf. [128, 130]).

An element $(x_0, x_0^*) \in X \times X^*$ is said to be *monotonically related* to the graph of S if

$$\langle y^* - x_0^*, y - x_0 \rangle \geq 0 \ \text{for all} \ (y, y^*) \in G(S).$$

One can show that a monotone operator S is maximal monotone if and only if the set of monotonically related elements to $G(S)$ is exactly $G(S)$.

To an arbitrary monotone operator $S : X \rightrightarrows X^*$ we associate the *Fitzpatrick function $\varphi_S : X \times X^* \to \overline{\mathbb{R}}$*, defined by

$$\varphi_S(x, x^*) = \sup\{\langle y^*, x \rangle + \langle x^*, y \rangle - \langle y^*, y \rangle : y^* \in S(y)\},$$

which is obviously convex and strong-weak* lower semicontinuous (it is even weak-weak* lower semicontinuous) in the corresponding topology on $X \times X^*$. Introduced by FITZPATRICK in 1988 (cf. [65]) and rediscovered after some years in [51, 100], it proved to be very important in the theory of maximal monotone operators, revealing important connections between convex analysis and monotone operators (see [5, 11, 27, 32, 33, 51, 98, 107–109, 113, 130, 132, 142, 151] and the references therein). Considering the function $c : X \times X^* \to \mathbb{R}$, $c(x, x^*) = \langle x^*, x \rangle$ for all $(x, x^*) \in X \times X^*$, we get the equality $\varphi_S(x, x^*) = c_S^*(x^*, x)$ for all $(x, x^*) \in X \times X^*$, where $c_S = c + \delta_{G(S)}$ and we are considering the natural injection $X \subseteq X^{**}$. The function $\psi_S = \mathrm{cl}_{\|\cdot\| \times \|\cdot\|_*}(\mathrm{co}\, c_S)$, where the closure is taken in the strong topology of $X \times X^*$, is well-linked to the Fitzpatrick function. Its properties were intensively studied in reflexive Banach spaces in [109] and in general Banach spaces in [51]. Let us mention that on $X \times X^*$ we have $\psi_S^{*^\top} = \varphi_S$ and, in the framework of reflexive Banach spaces the equality $\varphi_S^{*^\top} = \psi_S$ holds (cf. [51, Remark 5.4]). Let us recall the most important properties of the Fitzpatrick function.

Lemma 4.2 *(cf. [65]) Let $S : X \rightrightarrows X^*$ be a maximal monotone operator. Then*

(i) $\varphi_S(x, x^*) \geq \langle x^*, x \rangle$ *for all* $(x, x^*) \in X \times X^*$;

(ii) $G(S) = \{(x, x^*) \in X \times X^* : \varphi_S(x, x^*) = \langle x^*, x \rangle\}$.

Motivated by these properties of the Fitzpatrick function, the notion of *representative function* of a monotone operator was introduced and studied in the literature.

Definition 4.1 *For* $S : X \rightrightarrows X^*$ *a monotone operator, we call* representative function *of* S *a convex and lower semicontinuous function* $h_S : X \times X^* \to \overline{\mathbb{R}}$ *(in the strong topology of* $X \times X^*$*) fulfilling*

$$h_S \geq c \text{ and } G(S) \subseteq \{(x, x^*) \in X \times X^* : h_S(x, x^*) = \langle x^*, x \rangle\}.$$

We observe that if $G(S) \neq \emptyset$ (in particular if S is maximal monotone), then every representative function of S is proper. It follows immediately that the Fitzpatrick function associated to a maximal monotone operator is a representative function of the operator. The following proposition is a direct consequence of some results given in [51] (see also [94, Proposition 1.2 and Theorem 4.2 (1)]).

Proposition 4.1 *Let* $S : X \rightrightarrows X^*$ *be a maximal monotone operator and* h_S *be a representative function of* S. *Then*

(i) $\varphi_S \leq h_S \leq \psi_S$;

(ii) the canonical restriction of $h_S^{*\top}$ *to* $X \times X^*$ *is also a representative function of* S;

(iii) $\{(x, x^*) \in X \times X^* : h_S(x, x^*) = \langle x^*, x \rangle\} = \{(x, x^*) \in X \times X^* : h_S^{*\top}(x, x^*) = \langle x^*, x \rangle\} = G(S)$.

One can see by Proposition 4.1 that a convex and lower semicontinuous function $f : X \times X^* \to \overline{\mathbb{R}}$ is a representative function of the maximal monotone operator S if and only if $\varphi_S \leq f \leq \psi_S$. In particular, φ_S and ψ_S are representative functions of S and are the extremal elements of the family of representative functions associated to S

$$\mathcal{H}(S) = \left\{ h : X \times X^* \to \overline{\mathbb{R}} : \begin{array}{l} h \text{ is convex and lower semicontinuous, } h \geq c \text{ and} \\ G(S) \subseteq \{(x, x^*) \in X \times X^* : h(x, x^*) = \langle x^*, x \rangle\} \end{array} \right\},$$

which was introduced by BURACHIK AND SVAITER in [51].

Let us notice that if $f : X \to \overline{\mathbb{R}}$ is a proper, convex and lower semicontinuous function, then a representative function of the maximal monotone operator $\partial f : X \rightrightarrows X^*$ is the function $(x, x^*) \mapsto f(x) + f^*(x^*)$. This follows by the Young-Fenchel inequality and from the definition of the subdifferential of f. Moreover, according to [43, Theorem 3.1] (see also [108, Example 3]), if f is a sublinear and lower semicontinuous function, then the operator $\partial f : X \rightrightarrows X^*$ has a unique representative function, namely the function $(x, x^*) \mapsto f(x) + f^*(x^*)$.

If X is a Hilbert space, then there exists a unique representative function of the maximal monotone operator $\partial \delta_C : X \rightrightarrows X$, where C is a non-empty closed convex set in X. Indeed, by [6, Example 3.1], the Fitzpatrick function of $\partial \delta_C$ is $\varphi_{\partial \delta_C}(x, x^*) = \delta_C(x) + \delta_C^*(x^*)$. This implies that $\psi_{\partial \delta_C} = \varphi_{\partial \delta_C}^{*\top} = \varphi_{\partial \delta_C}$. As $f_{\partial \delta_C}$ is a representative function of $\partial \delta_C$ if and only if $\varphi_{\partial \delta_C} \leq f_{\partial \delta_C} \leq \psi_{\partial \delta_C}$, we get that the unique representative function is $(x, x^*) \mapsto \delta_C(x) + \delta_C^*(x^*)$.

For more on the properties of representative functions we refer to [11, 27, 51, 98, 113] and the references therein.

Remark 4.5 In many situations the representative functions are lower semicontinuous in the strong-weak* topology, as it is the case for example for the Fitzpatrick functions of monotone operators. As Proposition 4.1(ii) shows, for every representative function of a maximal monotone operator one obtains a corresponding representative function which is strong-weak* lower semicontinuous. Moreover, when $S = \partial f$, where $f : X \to \overline{\mathbb{R}}$ is a proper, convex and lower semicontinuous function, then the function $(x, x^*) \mapsto f(x) + f^*(x^*)$, which is a representative of ∂f, is lower semicontinuous in the strong-weak* topology. Hence, for $S : X \rightrightarrows X^*$ a monotone operator, it is natural to consider also the subfamily of $\mathcal{H}(S)$ formed by those representative functions of S which are lower semicontinuous with respect to the strong-weak* topology of $X \times X^*$. Let us notice that in general this is a proper subfamily (cf. [143, Remark 1]), while in the setting of reflexive Banach spaces it coincides with $\mathcal{H}(S)$. In Section 4.5 we will consider strong-weak* lower semicontinuous representative functions of maximal monotone operators.

Let us give the following maximality criteria valid in reflexive Banach spaces (cf. [52, Theorem 3.1] and [113, Proposition 2.1]; see also [128] for other maximality criteria in reflexive spaces). We refer to [94, Theorem 4.2] for a generalization of the next result to arbitrary Banach spaces.

Theorem 4.3 *(cf. [52,113]) Let X be a reflexive Banach space and $f : X \times X^* \to \overline{\mathbb{R}}$ a proper, convex and lower semicontinuous function such that $f \geq c$. Then the operator whose graph is the set $\{(x, x^*) \in X \times X^* : f(x, x^*) = \langle x^*, x \rangle\}$ is maximal monotone if and only if $f^{*\top} \geq c$.*

The following particular class of maximal monotone operators has been recently introduced in [94], being also studied in [144].

Definition 4.2 *An operator $S : X \rightrightarrows X^*$ is said to be* strongly-representable *whenever there exists a proper, convex and strong lower semicontinuous function $h : X \times X^* \to \overline{\mathbb{R}}$ such that*

$$h \geq c, h^*(x^*, x^{**}) \geq \langle x^{**}, x^* \rangle \; \forall (x^*, x^{**}) \in X^* \times X^{**}$$

and

$$G(S) = \{(x, x^*) \in X \times X^* : h(x, x^*) = \langle x^*, x \rangle\}.$$

In this case h is called a strong-representative *of S.*

If $S : X \rightrightarrows X^*$ is strongly-representable, then S is maximal monotone (see [94, Theorem 4.2] and [144, Theorem 8]) and φ_S is a strong-representative of S.

Remark 4.6 MARQUES ALVES AND SVAITER recently proved that the class of strongly-representable operators, the class of maximal monotone operators of type (NI) and the class of maximal monotone operators of Gossez type (D) coincide (cf. [95, Theorem 1.2] and [96, Theorem 4.4]).

The following definition of a family of enlargements associated to a monotone operator was introduced by SVAITER (cf. [134]).

Definition 4.3 *(cf. [134]) Let $S : X \rightrightarrows X^*$ be a monotone operator. Define $\mathbb{E}(S)$ as the family of multifunctions $E : \mathbb{R}_+ \times X \rightrightarrows X^*$ satisfying the following properties:*

(i) E is an enlargement of S, i.e.:

$$S(x) \subseteq E(\varepsilon, x) \text{ for all } \varepsilon \geq 0 \text{ and } x \in X;$$

(ii) E is non-decreasing, that is for all $x \in X$, $E(\varepsilon_1, x) \subseteq E(\varepsilon_2, x)$ provided that $\varepsilon_1 \leq \varepsilon_2$;

(iii) E satisfies the transportation formula: for every $(\varepsilon_1, x^1, v^1)$, $(\varepsilon_2, x^2, v^2) \in G(E)$ and for every $\lambda \in [0, 1]$ we have $(\varepsilon, x, v) \in G(E)$, where $\varepsilon := \lambda \varepsilon_1 + (1 - \lambda)\varepsilon_2 + \lambda(1 - \lambda)\langle v^1 - v^2, x^1 - x^2 \rangle$, $x := \lambda x^1 + (1 - \lambda)x^2$ and $v := \lambda v^1 + (1 - \lambda)v^2$.

A particular choice of $E \in \mathbb{E}(S)$ was considered in [46] and it has for $\varepsilon \geq 0$ and $x \in X$ the following definition

$$S^e(\varepsilon, x) = \{x^* \in X^* : \langle y^* - x^*, y - x \rangle \geq -\varepsilon \text{ for all } (y, y^*) \in G(S)\}.$$

Introduced in [46], this enlargement turned out to have some useful applications and properties similar to those of the ε-subdifferential (several properties like local boundedness, demiclosed graph, Lipschitz continuity, the Brøndsted-Rockafellar property were studied in [46, 50, 51, 134]). Notice that the Brøndsted-Rockafellar type property for the enlargement of a maximal monotone operator has been established for the first time in [140, Proposition 6.17] (see also [130, Theorem 29.9]). It is worth noting that $G\big(S^e(0, \cdot)\big)$ is exactly the set of the elements that are monotonically related to $G(S)$, hence the monotone operator S is maximal monotone if and only if $G(S) = G\big(S^e(0, \cdot)\big)$ (cf. [46, Proposition 2] and [119, Proposition 3.1]). In case S is maximal monotone, the operator S^e belongs to $\mathbb{E}_c(S)$ (the family of enlargements $E \in \mathbb{E}(S)$ such that $G(E)$ is closed with respect to the strong topology on $X \times X^*$) and in fact it is the *biggest element* of $\mathbb{E}_c(S)$ (cf. [134]). The enlargement S^e can be characterized via the Fitzpatrick function associated to S as follows: for $\varepsilon \geq 0$ and $x \in X$ we have

$$S^e(\varepsilon, x) = \{x^* \in X^* : \varphi_S(x, x^*) \leq \varepsilon + \langle x^*, x \rangle\}.$$

The family $\mathbb{E}_c(S)$ has also a *smallest element*, namely the enlargement S^{se} defined for all $(\varepsilon, x) \in \mathbb{R}_+ \times X$ by

$$S^{se}(\varepsilon, x) = \bigcap_{E \in \mathbb{E}_c(S)} E(\varepsilon, x).$$

Remark 4.7 Given $E : \mathbb{R}_+ \times X \rightrightarrows X^*$, we define the *closure* of E, $\overline{E} : \mathbb{R}_+ \times X \rightrightarrows X^*$ by (cf. [134])

$$\overline{E}(\varepsilon, x) := \{x^* \in X^* : (\varepsilon, x, x^*) \in \mathrm{cl}\big(G(E)\big)\} \ \forall (\varepsilon, x) \in \mathbb{R}_+ \times X.$$

We say that E is closed if $E = \overline{E}$. One can see that $G(\overline{E}) = \mathrm{cl}\big(G(E)\big)$. Consider in the following $S : X \rightrightarrows X^*$ a maximal monotone operator. The smallest element of the family $\mathbb{E}_c(S)$ was introduced in [51] by the following procedure. Define $M^S : \mathbb{R}_+ \times X \rightrightarrows X^*$, by

$$M^S(\varepsilon, x) := \bigcap_{E \in \mathbb{E}(S)} E(\varepsilon, x) \ \forall (\varepsilon, x) \in \mathbb{R}_+ \times X.$$

Then $\overline{M^S}$ is the smallest element of $\mathbb{E}_c(S)$ (cf. [51, Proposition 2.6]). In the following we show that $\overline{M^S} = S^{se}$. Indeed, from the definitions above one has $G(M^S) \subseteq G(S^{se})$, hence $G(\overline{M^S}) \subseteq G(S^{se})$ (one can prove that $G(S^{se})$ is closed). On the other hand, take $(\varepsilon, x) \in \mathbb{R}_+ \times X$ and $x^* \in S^{se}(\varepsilon, x)$ arbitrary. Then $x^* \in E(\varepsilon, x)$ for all $E \in \mathbb{E}_c(S)$, hence in particular (since $\overline{M^S} \in \mathbb{E}_c(S)$, cf. [51, Proposition 2.6]) $x^* \in \overline{M^S}(\varepsilon, x)$. Thus $G(S^{se}) \subseteq G(\overline{M^S})$. All together, we obtain $\overline{M^S} = S^{se}$. Let us notice that M^S is the smallest element of $\mathbb{E}(S)$ (cf. [134, Lemma 3.6]).

For an arbitrary representative function h_S one can consider the following enlargement of S (see [51, 53]): $S_{h_S} : \mathbb{R}_+ \times X \rightrightarrows X^*$,

$$S_{h_S}(\varepsilon, x) := \{x^* \in X^* : h_S(x, x^*) \leq \varepsilon + \langle x^*, x \rangle\} \text{ for all } (\varepsilon, x) \in \mathbb{R}_+ \times X.$$

Let us notice that for all $\varepsilon \geq 0$ the set $S_{h_S}(\varepsilon, x)$ is convex and closed (weak*-closed) if h_S is lower semicontinuous in the strong (strong-weak*) topology of $X \times X^*$. It follows immediately from the definitions above that $S_{\varphi_S} = S^e$. It was proved (see [51]) that for a maximal monotone operator S, $S_{h_S} \in \mathbb{E}_c(S)$ and actually (cf. [51, Theorem 3.6]) there exists a one-to-one correspondence between $\mathbb{E}_c(S)$ and the set $\mathcal{H}(S)$ (moreover, this correspondence is an isomorphism with respect to some suitable operations, see [53]). Hence, in case S is a maximal monotone operator, there exists a unique function belonging to $\mathcal{H}(S)$ such that $S^{se} = S_{h_S}$ and in fact $S^{se} = S_{\psi_S}$ (cf. [51, relation (35)], see also Remark 4.7 above). Further, for all $(\varepsilon, x) \in \mathbb{R}_+ \times X$ we have

$$S(x) \subseteq S^{se}(\varepsilon, x) = S_{\psi_S}(\varepsilon, x) \subseteq \begin{matrix} S_{h_S}(\varepsilon, x) \\ \\ S_{h_S^*}(\varepsilon, x) \end{matrix} \subseteq S_{\varphi_S}(\varepsilon, x) = S^e(\varepsilon, x),$$

where $S_{h_S^*}(\varepsilon, x) = \{x^* \in X^* : h_S^*(x^*, x) \leq \varepsilon + \langle x^*, x \rangle\}$, as well as $S(x) = S^{se}(0, x) = S_{h_S}(0, x) = S_{h_S^*}(0, x) = S^e(0, x)$.

Remark 4.8 If $S = \partial f$, where f is a proper, convex and lower semicontinuous function, then for all $\varepsilon \geq 0$ and all $x \in X$ we have

$$\partial f(x) \subseteq \partial_\varepsilon f(x) \subseteq \partial^\varepsilon f(x) := (\partial f)^e(\varepsilon, x),$$

and the inclusions can be strict (see [46, 99]). Moreover, taking $h : X \times X^* \to \overline{\mathbb{R}}$, $h(x, x^*) = f(x) + f^*(x^*)$ for all $(x, x^*) \in X \times X^*$, which is a representative function of ∂f, we see that $(\partial f)_h(\varepsilon, x) = (\partial f)_{h^*}(\varepsilon, x) = \partial_\varepsilon f(x)$ for all $\varepsilon \geq 0$ and all $x \in X$.

4.3 The ε-enlargement of the sum of two maximal monotone operators

We show that the necessary and sufficient regularity condition given in Section 4.1 for the bivariate infimal convolution formula delivers a closedness-type regularity condition which equivalently characterizes the ε-enlargement of the sum of two maximal monotone operators. Let us begin with the following result concerning the representability of the sum operator. In this section X is a Banach space.

Theorem 4.4 Let $S, T : X \rightrightarrows X^*$ be two maximal monotone operators with representative functions h_S and h_T, respectively, such that $\mathrm{pr}_X(\mathrm{dom}\, h_S) \cap \mathrm{pr}_X(\mathrm{dom}\, h_T) \neq \emptyset$ and consider the function $h : X \times X^* \to \overline{\mathbb{R}}$, $h(x, x^*) = (h_S \square_2 h_T)^*(x^*, x)$ for all $(x, x^*) \in X \times X^*$. If

$$\{(a^* + b^*, u^{**}, v^{**}, r) : h_S^*(a^*, u^{**}) + h_T^*(b^*, v^{**}) \leq r\} \text{ is closed regarding the set}$$
$$X^* \times \Delta_X \times \mathbb{R} \text{ in } (X^*, \omega(X^*, X)) \times (X^{**}, \omega(X^{**}, X^*)) \times (X^{**}, \omega(X^{**}, X^*)) \times \mathbb{R},$$

then h is a representative function of the monotone operator $S + T$. If, additionally, X is reflexive, then $S + T$ is a maximal monotone operator.

Proof. The function h is obviously convex and strong-weak* lower semicontinuous, hence lower semicontinuous in the strong topology of $X \times X^*$. Applying Corollary 4.3 we obtain $h(x, x^*) = (h_S^* \square_1 h_T^*)(x^*, x)$ and $h_S^* \square_1 h_T^*$ is exact at (x^*, x) for all

$(x^*, x) \in X^* \times X$. By using Proposition 4.1 we have for all $(x, x^*) \in X \times X^*$ that $h(x, x^*) = (h_S^* \Box_1 h_T^*)(x^*, x) = \inf\{h_S^*(u^*, x) + h_T^*(v^*, x) : u^*, v^* \in X^*, u^* + v^* = x^*\} \geq \inf\{\langle u^*, x \rangle + \langle v^*, x \rangle : u^*, v^* \in X^*, u^* + v^* = x^*\} = \langle x^*, x \rangle$, hence $h \geq c$.

It remains to show that $G(S + T) \subseteq \{(x, x^*) : h(x, x^*) = \langle x^*, x \rangle\}$. Take an arbitrary $(x, x^*) \in G(S + T)$. There exist $u^* \in S(x)$ and $v^* \in T(x)$ such that $x^* = u^* + v^*$. Employing once more Proposition 4.1 we obtain

$$\langle x^*, x \rangle \leq h(x, x^*) = (h_S^* \Box_1 h_T^*)(x^*, x)$$

$$\leq h_S^*(u^*, x) + h_T^*(v^*, x) = \langle u^*, x \rangle + \langle v^*, x \rangle = \langle x^*, x \rangle,$$

thus $G(S + T) \subseteq \{(x, x^*) : h(x, x^*) = \langle x^*, x \rangle\}$.

Actually, we prove that in this case

$$G(S + T) = \{(x, x^*) : h(x, x^*) = \langle x^*, x \rangle\}. \tag{4.7}$$

Take an arbitrary (x, x^*) such that $h(x, x^*) = \langle x^*, x \rangle$. Since we have that $h(x, x^*) = (h_S^* \Box_1 h_T^*)(x^*, x)$ and $h_S^* \Box_1 h_T^*$ is exact at (x^*, x), there exist $u^*, v^* \in X^*$, $u^* + v^* = x^*$ such that

$$h_S^*(u^*, x) + h_T^*(v^*, x) = \langle u^*, x \rangle + \langle v^*, x \rangle. \tag{4.8}$$

The function h_S and h_T being representative, from Proposition 4.1 we have $h_S^*(u^*, x) \geq \langle u^*, x \rangle$ and $h_T^*(v^*, x) \geq \langle v^*, x \rangle$, hence, in view of (4.8), the inequalities above must be fulfilled as equalities. Thus, by Proposition 4.1 we get $u^* \in S(x)$ and $v^* \in T(x)$, so $x^* = u^* + v^* \in S(x) + T(x) = (S + T)(x)$ and (4.7) is fulfilled.

Suppose now that X is a reflexive Banach space. Since in this case the weak* topology coincides with the weak topology (on X^*) and the weak closure of a convex set is exactly the strong closure of the same set, the regularity condition becomes

$\{(a^* + b^*, u, v, r) : h_S^*(a^*, u) + h_T^*(b^*, v) \leq r\}$ is closed regarding the subspace $X^* \times \Delta_X \times \mathbb{R}$ in the strong topology of $X^* \times X \times X \times \mathbb{R}$,

which is exactly the condition given in [27] for the maximal monotonicity of the operator $S + T$. However, we give in the following a different proof of this result.

Since h_S and h_T are representative functions we have $h_S \Box_2 h_T \geq c$. As the duality product is continuous with respect to the strong topology of $X \times X^*$, it follows $\text{cl}_{\|\cdot\| \times \|\cdot\|_*}(h_S \Box_2 h_T) \geq c$. Taking into account that the space X is reflexive and the functions h_S and h_T are convex, from the definition of h we obtain $h^{*\top} = \text{cl}_{\|\cdot\| \times w^*}(h_S \Box_2 h_T) = \text{cl}_{\|\cdot\| \times \|\cdot\|_*}(h_S \Box_2 h_T) \geq c$. The conclusion follows now by combining Theorem 4.3 with relation (4.7). $\qquad \Box$

Remark 4.9 In case of reflexive Banach spaces the condition given in the above theorem is the weakest one given so far which guarantees the maximality of the sum of two maximal monotone operators. Let us notice that a stronger closedness-type regularity condition is considered in [87]. We refer to [27, 32, 33] for a discussion regarding several other conditions given in the literature on this topic by means of generalized interiority notions. We mention here the *Rockafellar Condition*, namely $\text{int}\left(\text{dom}(S)\right) \cap \text{dom}(T) \neq \emptyset$ (cf. [125]), which is one of the oldest introduced in the literature for the maximal monotonicity of the sum of two maximal monotone operators in reflexive Banach spaces. In the nonreflexive case, it is still an open question whether this condition is sufficient. Some important steps in the study of maximal monotone operators in nonreflexive Banach spaces have been made by BAUSCHKE, WANG AND YAO (cf. [7]), BORWEIN (cf. [11]), MARQUES ALVES AND SVAITER (cf. [94–96]), SIMONS (cf. [130]), VOISEI (cf. [142]), Zagrodny (cf. [146]), respectively VOISEI AND ZĂLINESCU (cf. [144]).

Let us state now the main result of this section.

Theorem 4.5 *Let $S, T : X \rightrightarrows X^*$ be two maximal monotone operators with representative functions h_S and h_T, respectively, such that $\mathrm{pr}_X(\mathrm{dom}\, h_S) \cap \mathrm{pr}_X(\mathrm{dom}\, h_T) \neq \emptyset$ and consider again the function h defined as in the previous theorem. Then the following statements are equivalent:*

(i) $\{(a^* + b^*, u^{**}, v^{**}, r) : h_S^*(a^*, u^{**}) + h_T^*(b^*, v^{**}) \leq r\}$ is closed regarding the set $X^* \times \Delta_X \times \mathbb{R}$ in $(X^*, \omega(X^*, X)) \times (X^{**}, \omega(X^{**}, X^*)) \times (X^{**}, \omega(X^{**}, X^*)) \times \mathbb{R}$;

(ii) $(S + T)_h(\varepsilon, x) = \displaystyle\bigcup_{\substack{\varepsilon_1, \varepsilon_2 \geq 0 \\ \varepsilon_1 + \varepsilon_2 = \varepsilon}} \left(S_{h_S^*}(\varepsilon_1, x) + T_{h_T^*}(\varepsilon_2, x) \right)$ for all $\varepsilon \geq 0$ and $x \in X$;

where for every $x \in X$ and $\varepsilon \geq 0$, $(S + T)_h(\varepsilon, x) := \{x^* \in X^* : h(x, x^*) \leq \varepsilon + \langle x^*, x \rangle\}$.

Proof. Let us suppose that (i) is fulfilled and take $x \in X$ and $\varepsilon \geq 0$. We show first the inclusion

$$\bigcup_{\substack{\varepsilon_1, \varepsilon_2 \geq 0 \\ \varepsilon_1 + \varepsilon_2 = \varepsilon}} \left(S_{h_S^*}(\varepsilon_1, x) + T_{h_T^*}(\varepsilon_2, x) \right) \subseteq (S + T)_h(\varepsilon, x). \tag{4.9}$$

Indeed, take $\varepsilon_1, \varepsilon_2 \geq 0$, $\varepsilon_1 + \varepsilon_2 = \varepsilon$, $u^* \in S_{h_S^*}(\varepsilon_1, x)$ and $v^* \in T_{h_T^*}(\varepsilon_2, x)$. Then $h(x, u^* + v^*) = (h_S \square_2 h_T)^*(u^* + v^*, x) \leq (h_S^* \square_1 h_T^*)(u^* + v^*, x) \leq h_S^*(u^*, x) + h_T^*(v^*, x) \leq \varepsilon_1 + \langle u^*, x \rangle + \varepsilon_2 + \langle v^*, x \rangle = \varepsilon + \langle u^* + v^*, x \rangle$, hence $u^* + v^* \in (S+T)_h(\varepsilon, x)$, that is, the inclusion (4.9) is true. Let us mention that this inclusion is always fulfilled, as there is no need of (i) to prove (4.9).

However, to show the opposite inclusion, we use condition (i). Take $x^* \in (S + T)_h(\varepsilon, x)$. We have $(h_S \square_2 h_T)^*(x^*, x) \leq \varepsilon + \langle x^*, x \rangle$. By applying Corollary 4.3, we get $(h_S^* \square_1 h_T^*)(x^*, x) \leq \varepsilon + \langle x^*, x \rangle$ and the infimum in the definition of $(h_S^* \square_1 h_T^*)(x^*, x)$ is attained. Hence, there exist $u^*, v^* \in X^*$ such that $u^* + v^* = x^*$ and

$$h_S^*(u^*, x) + h_T^*(v^*, x) \leq \varepsilon + \langle u^*, x \rangle + \langle v^*, x \rangle. \tag{4.10}$$

Take $\varepsilon_1 := h_S^*(u^*, x) - \langle u^*, x \rangle$ and $\varepsilon_2 := \varepsilon - \varepsilon_1$. By using Proposition 4.1 and the inequality (4.10) we obtain $\varepsilon_1 \geq 0$ and $\varepsilon_2 \geq h_T^*(v^*, x) - \langle v^*, x \rangle \geq 0$. Thus $u^* \in S_{h_S^*}(\varepsilon_1, x)$ and $v^* \in T_{h_T^*}(\varepsilon_2, x)$, that is

$$x^* = u^* + v^* \in \bigcup_{\substack{\varepsilon_1, \varepsilon_2 \geq 0 \\ \varepsilon_1 + \varepsilon_2 = \varepsilon}} \left(S_{h_S^*}(\varepsilon_1, x) + T_{h_T^*}(\varepsilon_2, x) \right),$$

so (ii) is fulfilled.

Conversely, assume that (ii) is true. We start by proving that

$$h(x, x^*) \geq \langle x^*, x \rangle \text{ for all } (x, x^*) \in X \times X^*. \tag{4.11}$$

Let us suppose that there exists $(x_0, x_0^*) \in X \times X^*$ such that $h(x_0, x_0^*) \leq \langle x_0^*, x_0 \rangle$. By using the condition (ii) for $\varepsilon = 0$ we obtain $x_0^* \in (S + T)_h(0, x_0) = S_{h_S^*}(0, x_0) + T_{h_T^*}(0, x_0) = S(x_0) + T(x_0)$. Hence there exist $u_0^* \in S(x_0)$ and $v_0^* \in T(x_0)$ such that $x_0^* = u_0^* + v_0^*$. From Proposition 4.1 we obtain $h_S(x_0, u_0^*) = \langle u_0^*, x_0 \rangle$ and $h_T(x_0, v_0^*) = \langle v_0^*, x_0 \rangle$. Like in (4.6) we get

$$h(x_0, x_0^*) = \sup_{x \in X, u^*, v^* \in X^*} \{\langle x_0^*, x \rangle + \langle u^*, x_0 \rangle + \langle v^*, x_0 \rangle - h_S(x, u^*) - h_T(x, v^*)\}$$

$$\geq \langle x_0^*, x_0 \rangle + \langle u_0^*, x_0 \rangle + \langle v_0^*, x_0 \rangle - h_S(x_0, u_0^*) - h_T(x_0, v_0^*) = \langle x_0^*, x_0 \rangle,$$

thus (4. 11) is fulfilled.

In view of Corollary 4.3, it is sufficient to show that $h(x, x^*) = (h_S^* \square_1 h_T^*)(x^*, x)$ and $h_S^* \square_1 h_T^*$ is exact at (x^*, x) for all $(x^*, x) \in X^* \times X$. Take an arbitrary $(x^*, x) \in X^* \times X$. The inequality

$$h(x, x^*) \leq (h_S^* \square_1 h_T^*)(x^*, x) \tag{4. 12}$$

is always true. In case when $h(x, x^*) = +\infty$, there is nothing to be proved. The relation (4. 11) ensures $h(x, x^*) > -\infty$, so we may suppose that $h(x, x^*) \in \mathbb{R}$. Let us denote by $r := h(x, x^*)$. We have $h(x, x^*) = \langle x^*, x \rangle + (r - \langle x^*, x \rangle)$. With $\varepsilon := r - \langle x^*, x \rangle \geq 0$ (cf. (4. 11)), we obtain $x^* \in (S + T)_h(\varepsilon, x)$. Since (ii) is true, there exist $\varepsilon_1, \varepsilon_2 \geq 0$, $\varepsilon_1 + \varepsilon_2 = \varepsilon$ and $u^* \in S_{h_S^*}(\varepsilon_1, x)$ and $v^* \in T_{h_T^*}(\varepsilon_2, x)$, respectively, such that $x^* = u^* + v^*$. Further, adding the two inequalities

$$h_S^*(u^*, x) \leq \varepsilon_1 + \langle u^*, x \rangle$$

and

$$h_T^*(v^*, x) \leq \varepsilon_2 + \langle v^*, x \rangle$$

we obtain

$$h_S^*(u^*, x) + h_T^*(v^*, x) \leq \varepsilon_1 + \varepsilon_2 + \langle u^* + v^*, x \rangle = r = h(x, x^*),$$

hence, in view of (4. 12) we get $h(x, x^*) = h_S^*(u^*, x) + h_T^*(v^*, x) = (h_S^* \square_1 h_2^*)(x^*, x)$ and the proof is complete. $\qquad\square$

Remark 4.10 In view of Theorem 4.4, in case the condition (i) in the theorem above is fulfilled, then h is a representative function of the operator $S + T$, hence the notation $(S + T)_h(\varepsilon, x) := \{x^* \in X^* : h(x, x^*) \leq \varepsilon + \langle x^*, x \rangle\}$ is justified. Conversely, when the condition (ii) is true, then (i) is also fulfilled (see the proof above), thus also in this case the use of this notation makes sense.

One can give also a generalized interior-point regularity condition in order to guarantee the equality (ii) in the previous result. The following corollary is a direct consequence of Theorem 4.5 combined with Remark 4.4.

Corollary 4.4 *Let* $S, T : X \rightrightarrows X^*$ *be two maximal monotone operators with representative functions* h_S *and* h_T, *respectively, such that* $\mathrm{pr}_X(\mathrm{dom}\, h_S) \cap \mathrm{pr}_X(\mathrm{dom}\, h_T) \neq \emptyset$. *If the following condition is satisfied*

$$(CQ^{SZ}) \qquad\qquad 0 \in \mathrm{sqri}\left(\mathrm{pr}_X(\mathrm{dom}\, h_S) - \mathrm{pr}_X(\mathrm{dom}\, h_T)\right),$$

then $(S + T)_h(\varepsilon, x) = \bigcup_{\substack{\varepsilon_1, \varepsilon_2 \geq 0 \\ \varepsilon_1 + \varepsilon_2 = \varepsilon}} \left(S_{h_S^*}(\varepsilon_1, x) + T_{h_T^*}(\varepsilon_2, x)\right)$ *for all* $\varepsilon \geq 0$ *and* $x \in X$.

Remark 4.11 The condition (CQ^{SZ}) in Corollary 4.4 is only sufficient for the equality (ii) in Theorem 4.5, as it can be seen by taking $X = \mathbb{R}^2$, $S = \partial f$, $T = \partial g$, where f, g, h_S and h_T are defined as in Example 4.1.

We show in the following that a result stated in [47, Theorem 1] for ε-subdifferentials can be derived from Theorem 4.5.

Corollary 4.5 *Let* $f, g : X \to \overline{\mathbb{R}}$ *be proper, convex and lower semicontinuous such that* $\mathrm{dom}\, f \cap \mathrm{dom}\, g \neq \emptyset$. *The following statements are equivalent:*

(i) epi f^* + epi g^* *is closed in* $(X^*, \omega(X^*, X)) \times \mathbb{R}$;

(ii) $\partial_\varepsilon(f+g)(x) = \displaystyle\bigcup_{\substack{\varepsilon_1,\varepsilon_2 \geq 0 \\ \varepsilon_1+\varepsilon_2=\varepsilon}} \left(\partial_{\varepsilon_1}f(x) + \partial_{\varepsilon_2}g(x)\right)$ *for all $\varepsilon \geq 0$ and $x \in X$.*

Proof. Consider the functions $h_1, h_2 : X \times X^* \to \overline{\mathbb{R}}$ defined by $h_1(x, x^*) = f(x) + f^*(x^*)$ and $h_2(x, x^*) = g(x) + g^*(x^*)$ for all $(x, x^*) \in X \times X^*$. We have $h_1^*(x^*, x^{**}) = f^{**}(x^{**}) + f^*(x^*)$ and $h_2^*(x^*, x^{**}) = g^{**}(x^{**}) + g^*(x^*)$ for all $(x^*, x^{**}) \in X^* \times X^{**}$. Further, the condition $(h_1 \Box_2 h_2)^*(x^*, x) = (h_1^* \Box_1 h_2^*)(x^*, x)$ and $h_1^* \Box_1 h_2^*$ is exact at (x^*, x) for all $(x^*, x) \in X^* \times X$ is fulfilled if and only if $(f + g)^* = f^* \Box g^*$ and $f^* \Box g^*$ is exact. Applying Corollary 4.1 for $U = X^*$, (i) is fulfilled if and only if $(f + g)^* = f^* \Box g^*$ and $f^* \Box g^*$ is exact, which is equivalent to $(h_1 \Box_2 h_2)^*(x^*, x) = (h_1^* \Box_1 h_2^*)(x^*, x)$ and $h_1^* \Box_1 h_2^*$ is exact at (x^*, x) for all $(x^*, x) \in X^* \times X$. The later one is equivalent to (see Corollary 4.3)

$\{(a^* + b^*, u^{**}, v^{**}, r) : h_1^*(a^*, u^{**}) + h_2^*(b^*, v^{**}) \leq r\}$ is closed regarding the subspace $X^* \times \Delta_X \times \mathbb{R}$ in $(X^*, \omega(X^*, X)) \times (X^{**}, \omega(X^{**}, X^*)) \times (X^{**}, \omega(X^{**}, X^*)) \times \mathbb{R}$.

Since h_1 and h_2 are representative functions of the maximal monotone operators ∂f and ∂g, respectively, we obtain, by Theorem 4.5, applied to the operators $S = \partial f$ and $T = \partial g$, that (i) is fulfilled if and only if for all $\varepsilon \geq 0$ and all $x \in X$ the following equality holds

$$(\partial f + \partial g)_h(\varepsilon, x) = \bigcup_{\substack{\varepsilon_1,\varepsilon_2 \geq 0 \\ \varepsilon_1+\varepsilon_2=\varepsilon}} \left((\partial f)_{h_1^*}(\varepsilon_1, x) + (\partial g)_{h_2^*}(\varepsilon_2, x)\right),$$

where $h : X \times X^* \to \overline{\mathbb{R}}$, $h(x, x^*) = (h_1 \Box_2 h_2)^*(x^*, x) = (f + g)(x) + (f + g)^*(x^*)$ for all $(x, x^*) \in X \times X^*$. Taking into consideration that $(\partial f + \partial g)_h(\varepsilon, x) = \{x^* \in X^* : (f + g)(x) + (f + g)^*(x^*) \leq \varepsilon + \langle x^*, x \rangle\} = \partial_\varepsilon(f + g)(x)$ and $(\partial f)_{h_1^*}(\varepsilon_1, x) = \partial_{\varepsilon_1}f(x)$, respectively, $(\partial g)_{h_2^*}(\varepsilon_2, x) = \partial_{\varepsilon_2}g(x)$ (cf. Remark 4.8), we get the desired conclusion. $\square$

Remark 4.12 (a) The equivalence in Corollary 4.5 holds also in the framework of separated locally convex spaces (cf. [37, Theorem 5]). The direct implication is shown in [74, Theorem 2.1]. Sufficient conditions which guarantee the equality in Corollary 4.5(ii) can be found in [149, Theorem 2.8.7].

(b) In reflexive Banach spaces one can deduce the equivalence in Corollary 4.5 by using the results presented in [53] for enlargements of monotone operators (see [53, Theorem 6.9]).

(c) Following the approach presented above, one can give a similar result to Theorem 4.5, where, instead of $S + T$ one can consider the operator $S + A^*TA$, where $S : X \rightrightarrows X^*$ and $T : Y \rightrightarrows Y^*$ are maximal monotone operators, X, Y are Banach spaces and $A : X \to Y$ is a continuous linear operator.

4.4 A characterization of the maximal monotone operators which are fully enlargeable by S^{se}

In this section we give an answer to an open problem posed by BURACHIK AND IUSEM in [44]. Let us recall first the notion of *full enlargeability* introduced in [44]. We consider X a Banach space.

Definition 4.4 *(cf. [44]) Let $S : X \rightrightarrows X^*$ be a maximal monotone operator and consider an element $E \in \mathbb{E}(S)$. We say that*

(i) *the enlargement E fully enlarges S at the point $x \in D(S)$ if and only if for all $\varepsilon > 0$ there exists $\delta = \delta(x, \varepsilon) > 0$ such that $S(x) + B(0, \delta) \subseteq E(\varepsilon, x)$ $(B(0, \delta)$ is the closed ball centered at the origin with radius $\delta)$;*

(ii) *E is a full enlargement of S when property (i) holds for all $x \in D(S)$.*

The operators which are fully enlargeable by S^e are characterized in [44, Theorem 3.2]. The question posed in [44] concerning the characterization of the maximal monotone operators that are fully enlargeable by S^{se} was left as an open problem. We give below an answer to this question. Actually we provide a more general result, namely a characterization of the maximal monotone operators S which are fully enlargeable by S_{h_S}, where h_S is an arbitrary representative function of S. To this end for an operator $S : X \rightrightarrows X^*$ we introduce, as in [44], the function $\beta_S : X \times X^* \to \overline{\mathbb{R}}$, $\beta_S(x, x^*) = h_S(x, x^*) - \langle x, x^* \rangle$ and for $x^* \in X^*$ and $U \subseteq X^*$ consider the *metric distance* from x^* to U, that is $d(x^*, U) = \inf_{u^* \in U} \|u^* - x^*\|$.

Theorem 4.6 *Let $S : X \rightrightarrows X^*$ be a maximal monotone operator and h_S be a representative function of S. Then the following statements are equivalent:*

(i) *S_{h_S} is a full enlargement of S;*

(ii) *for all $x \in D(S)$, $h_S(x, \cdot)$ is uniformly continuous on $S(x)$.*

Proof. We give first the proof of the implication $(i) \Rightarrow (ii)$, which is similar to the proof of implication $(a) \Rightarrow (b)$ in [44, Theorem 3.2]. Let be $x \in D(S)$. Taking into consideration the definition of the function β_S, the uniform continuity of $h_S(x, \cdot)$ is equivalent to the uniform continuity of $\beta_S(x, \cdot)$. For $x^* \in S(x)$ we fix $\varepsilon > 0$ and consider $\delta > 0$ such that $S(x) + B(0, \delta) \subseteq S_{h_S}(\varepsilon, x)$, which exists by the definition of full enlargeability. Take $y^* \in X^*$ such that $d(y^*, S(x)) < \delta$. Consequently, there exists $u^* \in S(x)$ such that $y^* - u^* \in B(0, \delta)$. Hence $y^* = u^* + (y^* - u^*) \in S(x) + B(0, \delta) \subseteq S_{h_S}(\varepsilon, x)$, that is $\beta_S(x, y^*) \leq \varepsilon$. We obtain $|\beta_S(x, y^*) - \beta_S(x, x^*)| = \beta_S(x, y^*) \leq \varepsilon$ for all $x^* \in S(x)$. As δ depends only on x and ε, (ii) holds.

Assume now that (ii) holds and fix $x \in D(S)$ and $\varepsilon > 0$. Since the function $\beta_S(x, \cdot)$ is uniformly continuous on $S(x)$, there exists $\delta > 0$ (which depends on x and ε) fulfilling

$$\beta_S(x, y^*) \leq \varepsilon, \text{ for all } y^* \in X^* \text{ such that } d(y^*, S(x)) < \delta. \qquad (4.\,13)$$

We claim that for $\overline{\delta} := (1/2)\delta$ we have $S(x) + B(0, \overline{\delta}) \subseteq S_{h_S}(\varepsilon, x)$. Indeed, take $x^* \in S(x)$ and $v^* \in B(0, \overline{\delta})$. Then $d(x^* + v^*, S(x)) = \inf_{u^* \in S(x)} \|x^* + v^* - u^*\| \leq \|v^*\| \leq \overline{\delta} < \delta$. Combining this inequality with $(4.\,13)$ we get $\beta_S(x, x^* + v^*) \leq \varepsilon$, which is nothing else than $x^* + v^* \in S_{h_S}(\varepsilon, x)$ and the claim is proved. Hence (i) is fulfilled and the proof is complete. $\qquad \square$

Remark 4.13 By taking $h_S = \varphi_S$ in Theorem 4.6 we obtain exactly the equivalence $(a) \Leftrightarrow (b)$ in [44, Theorem 3.2]. In this case a further equivalent characterization of full enlargeability of S by S^e can be given (see [44, Theorem 3.2 (c)]).

By taking in the previous result $h_S = \psi_S$ we obtain a characterization of the maximal monotone operators which are fully enlargeable by S^{se} (remember that $S^{se} = S_{\psi_S}$, see Section 4.2).

Corollary 4.6 *Let $S : X \rightrightarrows X^*$ be a maximal monotone operator. Then the following statements are equivalent:*

(i) *S^{se} is a full enlargement of S;*

(ii) *for all $x \in D(S)$, $\psi_S(x, \cdot)$ is uniformly continuous on $S(x)$.*

4.5   Guaranteeing the weak*-closedness of the set $S_{h_S}(\varepsilon_1, x) + T_{h_T}(\varepsilon_2, x)$

In this section we provide a weak *generalized interior-point regularity condition* which ensures that for $S, T : X \rightrightarrows X^*$ maximal monotone operators with strong-weak* lower semicontinuous representative functions h_S and h_T, respectively, the set $S_{h_S}(\varepsilon_1, x) + T_{h_T}(\varepsilon_2, x)$ is weak*-closed, for all $\varepsilon_1, \varepsilon_2 \geq 0$ and all $x \in X$. Some comments concerning similar results given in the literature are also made.

In the following we assume that the Banach space X is endowed with the strong topology, while its topological dual X^* is endowed with the weak* topology $\omega(X^*, X)$. Thus for a given function $f : X^* \to \overline{\mathbb{R}}$ its conjugate function $f^* : X \to \overline{\mathbb{R}}$ is defined by $f^*(x) = \sup_{x^* \in X^*} \{ \langle x^*, x \rangle - f(x^*) \}$ for all $x \in X$, while for $h : X \times X^* \to \overline{\mathbb{R}}$, $h^* : X^* \times X \to \overline{\mathbb{R}}$ is defined as $h^*(z^*, z) = \sup_{x \in X, x^* \in X^*} \{ \langle z^*, x \rangle + \langle x^*, z \rangle - h(x, x^*) \}$ for all $(z^*, z) \in X^* \times X$. We prove first a preliminary result which will be useful in the proof of the main theorem of this section and which is of its own interest.

Theorem 4.7 *Let $f, g : X^* \to \overline{\mathbb{R}}$ be two proper, convex and weak* lower semicontinuous functions such that $0 \in \operatorname{dom} f^* \cap \operatorname{dom} g^*$ and consider the sets $F = \{ x^* \in X^* : f(x^*) \leq 0 \}$ and $G = \{ x^* \in X^* : g(x^*) \leq 0 \}$. If $0 \in \operatorname{sqri}(\operatorname{dom} f^* - \operatorname{dom} g^*)$, then $F + G$ is weak*-closed.*

Proof. The sets F and G are both convex and weak*-closed. If $F + G$ is empty, then there is nothing to be proved. Therefore we assume that $F + G$ is a non-empty set and consider $\sigma_F, \sigma_G : X \to \overline{\mathbb{R}}$ the *support functions* of F and G defined by $\sigma_F(x) = \sup_{x^* \in F} \langle x^*, x \rangle$ and $\sigma_G(x) = \sup_{x^* \in G} \langle x^*, x \rangle$ for all $x \in X$. Both functions σ_F and σ_G are proper, convex and lower semicontinuous (both in $\omega(X, X^*)$ and in the strong topology) and $0 \in \operatorname{dom} \sigma_F \cap \operatorname{dom} \sigma_G$. Therefore, whenever the condition

$$0 \in \operatorname{sqri}(\operatorname{dom} \sigma_F - \operatorname{dom} \sigma_G) \tag{4.14}$$

is fulfilled, one has (cf. [149, Theorem 2.8.7 *(vii)*]) that

$$\sigma^*_{F+G} = (\sigma_F + \sigma_G)^* = \sigma^*_F \square \sigma^*_G = \delta_F \square \delta_G = \delta_{F+G}$$

and this guarantees that the set $F + G$ is weak*-closed.

To obtain the conclusion we show that the condition $0 \in \operatorname{sqri}(\operatorname{dom} f^* - \operatorname{dom} g^*)$ from the hypotheses secures (4.14). To this aim we consider the function $h : X \to \overline{\mathbb{R}}$ defined by $h = \inf_{\lambda > 0} (\lambda f)^*$. Since the function

$$(x, \lambda) \mapsto \begin{cases} (\lambda f)^*(x), & \text{if } \lambda > 0, \\ +\infty, & \text{otherwise} \end{cases}$$

is convex (on $X \times \mathbb{R}$), one has that h is convex, too. Moreover, from the definition of the function h we obtain that $h(x) \geq \langle x^*, x \rangle$ for all $x \in X$ and all $x^* \in F$. Hence $\operatorname{cl} h \geq \sigma_F > -\infty$ and consequently $\operatorname{cl} h = h^{**}$. We claim that $h^{**} = \sigma_F$. By using that f is proper, convex and weak* lower semicontinuous, we get for all $x^* \in X^*$

$$h^*(x^*) = \sup_{x \in X} \{ \langle x^*, x \rangle - \inf_{\lambda > 0} (\lambda f)^*(x) \} = \sup_{\substack{x \in X \\ \lambda > 0}} \{ \langle x^*, x \rangle - (\lambda f)^*(x) \}$$

$$= \sup_{\lambda > 0} \left\{ \sup_{x \in X} \{ \langle x^*, x \rangle - (\lambda f)^*(x) \} \right\} = \sup_{\lambda > 0} (\lambda f)^{**}(x^*) = \sup_{\lambda > 0} \{ \lambda f(x^*) \} = \delta_F(x^*).$$

Hence $h^* = \delta_F$ and taking the conjugates we obtain $h^{**} = \sigma_F$. All together we get $\operatorname{cl} h = \sigma_F$, thus $\operatorname{dom} h \subseteq \operatorname{dom} \sigma_F \subseteq \operatorname{cl}(\operatorname{dom} h)$. On the other hand, $x \in \operatorname{dom} h$ if and only if there exists $\lambda > 0$ such that $(\lambda f)^*(x) = \lambda f^*((1/\lambda)x) < +\infty$. This is further equivalent to the existence of $\lambda > 0$ such that $x \in \lambda \operatorname{dom} f^*$ or, in other words, to $x \in \cup_{\lambda > 0} \lambda \operatorname{dom} f^*$. Consequently, $\operatorname{dom} h = \cup_{\lambda > 0} \lambda \operatorname{dom} f^*$, that is

$$\cup_{\lambda > 0} \lambda \operatorname{dom} f^* \subseteq \operatorname{dom} \sigma_F \subseteq \operatorname{cl} \left(\cup_{\lambda > 0} \lambda \operatorname{dom} f^* \right).$$

Similarly one can prove that

$$\cup_{\lambda > 0} \lambda \operatorname{dom} g^* \subseteq \operatorname{dom} \sigma_G \subseteq \operatorname{cl} \left(\cup_{\lambda > 0} \lambda \operatorname{dom} g^* \right).$$

On the one hand it holds

$$\underset{\lambda > 0}{\cup} \lambda(\operatorname{dom} f^* - \operatorname{dom} g^*) \subseteq \underset{\lambda > 0}{\cup} \lambda \operatorname{dom} f^* - \underset{\mu > 0}{\cup} \mu \operatorname{dom} g^* \subseteq \operatorname{dom} \sigma_F - \operatorname{dom} \sigma_G$$

and, on the other hand, that

$$\operatorname{dom} \sigma_F - \operatorname{dom} \sigma_G \subseteq \operatorname{cl} \left(\underset{\lambda > 0}{\cup} \lambda \operatorname{dom} f^* \right) - \operatorname{cl} \left(\underset{\mu > 0}{\cup} \mu \operatorname{dom} g^* \right)$$

$$\subseteq \operatorname{cl} \left(\underset{\lambda > 0}{\cup} \lambda \operatorname{dom} f^* - \underset{\mu > 0}{\cup} \mu \operatorname{dom} g^* \right).$$

For $\lambda, \mu > 0$, $x \in \operatorname{dom} f^*$ and $y \in \operatorname{dom} g^*$, using that $0 \in \operatorname{dom} f^* \cap \operatorname{dom} g^*$ and the convexity of the sets $\operatorname{dom} f^*$ and $\operatorname{dom} g^*$, we have

$$\lambda x - \mu y = (\lambda + \mu)\left(\lambda/(\lambda + \mu)(x - 0) + \mu/(\lambda + \mu)(0 - y)\right) \in (\lambda + \mu)(\operatorname{dom} f^* - \operatorname{dom} g^*)$$

$$\subseteq \cup_{\lambda > 0} \lambda(\operatorname{dom} f^* - \operatorname{dom} g^*).$$

Thus

$$\underset{\lambda > 0}{\cup} \lambda(\operatorname{dom} f^* - \operatorname{dom} g^*) \subseteq \operatorname{dom} \sigma_F - \operatorname{dom} \sigma_G \subseteq \operatorname{cl} \left(\underset{\lambda > 0}{\cup} \lambda\left(\operatorname{dom} f^* - \operatorname{dom} g^* \right) \right)$$

and from here

$$\underset{\lambda > 0}{\cup} \lambda(\operatorname{dom} f^* - \operatorname{dom} g^*) \subseteq \underset{\lambda > 0}{\cup} \lambda(\operatorname{dom} \sigma_F - \operatorname{dom} \sigma_G) \subseteq \operatorname{cl} \left(\underset{\lambda > 0}{\cup} \lambda\left(\operatorname{dom} f^* - \operatorname{dom} g^* \right) \right).$$

The hypotheses ensure that $\cup_{\lambda > 0} \lambda(\operatorname{dom} f^* - \operatorname{dom} g^*)$ is a closed linear subspace of X and this means that the inclusions in the relation above are fulfilled as equalities. This has as consequence the fact that $\cup_{\lambda > 0} \lambda(\operatorname{dom} \sigma_F - \operatorname{dom} \sigma_G)$ is a closed linear subspace of X too, or, equivalently, $0 \in \operatorname{sqri}(\operatorname{dom} \sigma_F - \operatorname{dom} \sigma_G)$, which completes the proof. $\quad\square$

Remark 4.14 The function h was introduced in the above proof in order to show that in case $F \neq \emptyset$ we have $\operatorname{cl}(\operatorname{dom} \sigma_F) = \operatorname{cl} \left(\cup_{\lambda > 0} \lambda \operatorname{dom} f^* \right)$. Let us give an alternative proof of this fact which relies on some techniques of *asymptotic analysis*. Recall that for $A \subseteq X$ we denote by A_∞ its *recession cone* and for $u : X \to \overline{\mathbb{R}}$, the function $u_\infty : X \to \overline{\mathbb{R}}$, whose epigraph is $(\operatorname{epi} u)_\infty$ is called its *recession function* (sometimes called *asymptotic function*). The function f being proper, convex and weak* lower semicontinuous, we get $f_\infty = \sigma_{\operatorname{dom} f^*}$ (cf. [149, Exercise 2.23]) and $F_\infty = \{x^* \in X^* : f_\infty(x^*) \leq 0\}$ (cf. [149, page 74, relation (2.29)]). We have

$$F_\infty = \{x^* \in X^* : \sigma_{\operatorname{dom} f^*}(x^*) \leq 0\}$$

$$= \{x^* \in X^* : \langle x^*, y \rangle \leq 0 \; \forall y \in \operatorname{dom} f^*\} = N_{\operatorname{dom} f^*}(0).$$

Since f^* is convex and $0 \in \operatorname{dom} f^*$, $\left(N_{\operatorname{dom} f^*}(0)\right)^- = \operatorname{cl}\left(\cup_{\lambda>0} \lambda \operatorname{dom} f^*\right)$. Moreover, as F is convex and weak*-closed, we have (cf. [145, page 142], see also [73, relation (7)] or [54, page 259, relation (A.2)])

$$\operatorname{cl}(\operatorname{dom} \sigma_F) = \left[\left(\operatorname{cl}(\operatorname{co}(F))\right)_\infty\right]^- = F_\infty^-.$$

All together the desired conclusion follows. For more on asymptotic analysis we refer to the monograph of AUSLENDER AND TEBOULLE [3].

We come now to the proof of the main result of this section.

Theorem 4.8 *Let* $S, T : X \rightrightarrows X^*$ *be two maximal monotone operators with representative functions* h_S *and* h_T, *respectively, which are lower semicontinuous with respect to the strong-weak* topology of* $X \times X^*$. *If*

$$0 \in \operatorname{sqri}\left(\operatorname{pr}_X(\operatorname{dom} h_S^*) - \operatorname{pr}_X(\operatorname{dom} h_T^*)\right),$$

then for all $\varepsilon_1, \varepsilon_2 \geq 0$ *and all* $x \in X$ *the set* $S_{h_S}(\varepsilon_1, x) + T_{h_T}(\varepsilon_2, x)$ *is weak*-closed.*

Proof. Let $\varepsilon_1, \varepsilon_2 \geq 0$ and $x \in X$ be fixed. Assume that $S_{h_S}(\varepsilon_1, x) + T_{h_T}(\varepsilon_2, x)$ is non-empty. Thus $x \in \operatorname{pr}_X(\operatorname{dom} h_S) \cap \operatorname{pr}_X(\operatorname{dom} h_T)$. Consider the functions $f, g : X^* \to \overline{\mathbb{R}}$ defined by $f(x^*) = h_S(x, x^*) - \langle x^*, x \rangle - \varepsilon_1$ and $g(x^*) = h_T(x, x^*) - \langle x^*, x \rangle - \varepsilon_2$ for all $x^* \in X^*$. The functions f and g are proper, convex and weak* lower semicontinuous. Since $\inf_{x^* \in X^*} f(x^*) \geq -\varepsilon_1 > -\infty$ and $\inf_{x^* \in X^*} g(x^*) \geq -\varepsilon_2 > -\infty$, it yields $0 \in \operatorname{dom} f^* \cap \operatorname{dom} g^*$. Moreover, $S_{h_S}(\varepsilon_1, x) = \{x^* \in X^* : f(x^*) \leq 0\}$ and $T_{h_T}(\varepsilon_2, x) = \{x^* \in X^* : g(x^*) \leq 0\}$.

We apply Lemma 4.1(ii) (remember that on X^* we consider the weak* topology, hence $X \times X^*$ and $X^* \times X$ are in duality) and obtain

$$\operatorname{pr}_X(\operatorname{dom} h_S^*) \subseteq \operatorname{dom}(h_S(x, \cdot))^* \subseteq \operatorname{cl}_{w(X, X^*)}\left(\operatorname{pr}_X(\operatorname{dom} h_S^*)\right) = \operatorname{cl}_{\|\cdot\|}\left(\operatorname{pr}_X(\operatorname{dom} h_S^*)\right)$$

and, similarly,

$$\operatorname{pr}_X(\operatorname{dom} h_T^*) \subseteq \operatorname{dom}(h_T(x, \cdot))^* \subseteq \operatorname{cl}\left(\operatorname{pr}_X(\operatorname{dom} h_T^*)\right).$$

Consequently,

$$\operatorname{pr}_X(\operatorname{dom} h_S^*) - \operatorname{pr}_X(\operatorname{dom} h_T^*) \subseteq \operatorname{dom}(h_S(x, \cdot))^* - \operatorname{dom}(h_T(x, \cdot))^*$$

$$\subseteq \operatorname{cl}\left(\operatorname{pr}_X(\operatorname{dom} h_S^*)\right) - \operatorname{cl}\left(\operatorname{pr}_X(\operatorname{dom} h_T^*)\right) \subseteq \operatorname{cl}\left(\operatorname{pr}_X(\operatorname{dom} h_S^*) - \operatorname{pr}_X(\operatorname{dom} h_T^*)\right)$$

and from here one has

$$\cup_{\lambda>0} \lambda\left(\operatorname{pr}_X(\operatorname{dom} h_S^*) - \operatorname{pr}_X(\operatorname{dom} h_T^*)\right) \subseteq \cup_{\lambda>0} \lambda\left(\operatorname{dom}(h_S(x, \cdot))^* - \operatorname{dom}(h_T(x, \cdot))^*\right)$$

$$\subseteq \operatorname{cl}\left(\cup_{\lambda>0} \lambda\left(\operatorname{pr}_X(\operatorname{dom} h_S^*) - \operatorname{pr}_X(\operatorname{dom} h_T^*)\right)\right).$$

The hypotheses guarantee that the set $\cup_{\lambda>0}\lambda\left(\operatorname{pr}_X(\operatorname{dom} h_S^*) - \operatorname{pr}_X(\operatorname{dom} h_T^*)\right)$ is a closed linear subspace of X. Therefore $\cup_{\lambda>0} \lambda\left(\operatorname{dom}(h_S(x, \cdot))^* - \operatorname{dom}(h_T(x, \cdot))^*\right)$ is a closed linear subspace of X too, or, equivalently, $0 \in \operatorname{sqri}\left(\operatorname{dom}(h_S(x, \cdot))^* - \operatorname{dom}(h_T(x, \cdot))^*\right)$.

For all $u \in X$ we have $f^*(u) = \sup_{x^* \in X^*}\{\langle x^*, u + x \rangle - h_S(x, x^*)\} + \varepsilon_1 = (h_S(x, \cdot))^*(u + x) + \varepsilon_1$ and therefore $\operatorname{dom} f^* = \operatorname{dom}(h_S(x, \cdot))^* - x$. Analogously, we obtain $\operatorname{dom} g^* = \operatorname{dom}(h_T(x, \cdot))^* - x$, which implies that $0 \in \operatorname{sqri}(\operatorname{dom} f^* - \operatorname{dom} g^*)$. Finally, Theorem 4.7 guarantees that the set $S_{h_S}(\varepsilon_1, x) + T_{h_T}(\varepsilon_2, x)$ is weak*-closed. $\square$

The following theorem follows as a direct consequence of the result above by considering as representative functions of S and T their *Fitzpatrick functions*.

Theorem 4.9 *Let $S, T : X \rightrightarrows X^*$ be two maximal monotone operators. If*

$$0 \in \mathrm{sqri} \left(\mathrm{pr}_X(\mathrm{dom}\, \varphi_S^*) - \mathrm{pr}_X(\mathrm{dom}\, \varphi_T^*) \right),$$

then for all $\varepsilon_1, \varepsilon_2 \geq 0$ and all $x \in X$ the set $S^e(\varepsilon_1, x) + T^e(\varepsilon_2, x)$ is weak*-closed.

In the following we give another regularity condition which is sufficient for obtaining the same conclusion like in the theorem above, this time involving the domains of the two operators. To this end we recall a result stated in [132, Lemma 5.3], the proof of which uses techniques taken from [128, pp. 57–62 and pp. 87–88], in case X is a reflexive Banach space and the representative functions are exactly the Fitzpatrick functions. It can be proved in an analogous way (by using Proposition 4.1) that the result remains valid in a general Banach space and when considering arbitrary representative functions.

Lemma 4.3 *Let $S, T : X \rightrightarrows X^*$ be two maximal monotone operators with representative functions h_S and h_T, respectively. The following statements are true*

(a) *If F is a closed subspace of X, $w \in X$ and $D(S) \subseteq F + w$, then $\mathrm{pr}_X(\mathrm{dom}\, h_S) \subseteq F + w$;*

(b) *$\bigcup_{\lambda > 0} \lambda \left(\mathrm{pr}_X(\mathrm{dom}\, h_S) - \mathrm{pr}_X(\mathrm{dom}\, h_T) \right) \subseteq \mathrm{cl} \left(\mathrm{lin} \left(D(S) - D(T) \right) \right)$.*

Remark 4.15 It follows easily from Proposition 4.1 and Lemma 4.3 that for $S, T : X \rightrightarrows X^*$ maximal monotone operators the following inclusions hold:

$$\bigcup_{\lambda > 0} \lambda \left(D(S) - D(T) \right) \subseteq \bigcup_{\lambda > 0} \lambda \left(\mathrm{co}(D(S)) - \mathrm{co}(D(T)) \right)$$

$$\subseteq \bigcup_{\lambda > 0} \lambda \left(\mathrm{pr}_X(\mathrm{dom}\, \varphi_S^*) - \mathrm{pr}_X(\mathrm{dom}\, \varphi_T^*) \right) \subseteq \bigcup_{\lambda > 0} \lambda \left(\mathrm{pr}_X(\mathrm{dom}\, \varphi_S) - \mathrm{pr}_X(\mathrm{dom}\, \varphi_T) \right)$$

$$\subseteq \mathrm{cl} \left(\mathrm{lin} \left(D(S) - D(T) \right) \right) \subseteq \mathrm{cl} \left(\mathrm{lin} \left(\mathrm{co}(D(S)) - \mathrm{co}(D(T)) \right) \right)$$

$$\subseteq \mathrm{cl} \left(\mathrm{lin} \left(\mathrm{pr}_X(\mathrm{dom}\, \varphi_S^*) - \mathrm{pr}_X(\mathrm{dom}\, \varphi_T^*) \right) \right)$$

$$\subseteq \mathrm{cl} \left(\mathrm{lin} \left(\mathrm{pr}_X(\mathrm{dom}\, \varphi_S) - \mathrm{pr}_X(\mathrm{dom}\, \varphi_T) \right) \right) \subseteq \mathrm{cl} \left(\mathrm{lin} \left(D(S) - D(T) \right) \right),$$

thus

$$\mathrm{cl} \left(\mathrm{lin} \left(D(S) - D(T) \right) \right) = \mathrm{cl} \left(\mathrm{lin} \left(\mathrm{co}(D(S)) - \mathrm{co}(D(T)) \right) \right) =$$

$$\mathrm{cl} \left(\mathrm{lin} \left(\mathrm{pr}_X(\mathrm{dom}\, \varphi_S^*) - \mathrm{pr}_X(\mathrm{dom}\, \varphi_T^*) \right) \right) = \mathrm{cl} \left(\mathrm{lin} \left(\mathrm{pr}_X(\mathrm{dom}\, \varphi_S) - \mathrm{pr}_X(\mathrm{dom}\, \varphi_T) \right) \right).$$

The remark above allows us to formulate the following result.

Theorem 4.10 *Let $S, T : X \rightrightarrows X^*$ be two maximal monotone operators. If*

$$0 \in \mathrm{sqri} \left(\mathrm{co}(D(S)) - \mathrm{co}(D(T)) \right),$$

then for all $\varepsilon_1, \varepsilon_2 \geq 0$ and all $x \in X$ the set $S^e(\varepsilon_1, x) + T^e(\varepsilon_2, x)$ is weak*-closed.

For a particular instance of Theorem 4.10, when $\varepsilon_1 = \varepsilon_2 = 0$ and the stronger regularity condition $0 \in \mathrm{core} \left(\mathrm{co}(D(S)) - \mathrm{co}(D(T)) \right)$ is considered, we refer the reader to [141, Corollary 2.3].

Remark 4.16 In case X is a reflexive Banach space the generalized interior-point regularity conditions stated in Theorem 4.9 and Theorem 4.10 for the weak*-closedness of the set $S^e(\varepsilon_1, x) + T^e(\varepsilon_2, x)$, when $\varepsilon_1, \varepsilon_2 \geq 0$ and $x \in X$, are equivalent. More than that, they are further equivalent to (see [151])

$$0 \in \text{sqri}\left(D(S) - D(T)\right)$$

and to

$$0 \in \text{sqri}\left(\text{pr}_X(\text{dom } \varphi_S) - \text{pr}_X(\text{dom } \varphi_T)\right).$$

In case X is a general Banach space and the operators S and T are *strongly-representable*, then whenever $0 \in \text{sqri}\left(\text{pr}_X(\text{dom } \varphi_S) - \text{pr}_X(\text{dom } \varphi_T)\right)$ or, equivalently (see [144, Theorem 16]), $0 \in \text{sqri}\left(\text{co}(D(S)) - \text{co}(D(T))\right)$, or $0 \in \text{sqri}\left(D(S) - D(T)\right)$, then we also have that for $\varepsilon_1, \varepsilon_2 \geq 0$ and $x \in X$ the set $S^e(\varepsilon_1, x) + T^e(\varepsilon_2, x)$ is weak*-closed.

Remark 4.17 By using tools from the functional analysis, GARCÍA, LASSONDE AND REVALSKI proved in [67, Theorem 3.7] that in case X is a Banach space and $S, T : X \rightrightarrows X^*$ are two maximal monotone operators which satisfy the condition $0 \in \text{core}\left(\text{pr}_X(\text{dom } \varphi_S) - \text{pr}_X(\text{dom } \varphi_T)\right)$, one has for all $\varepsilon \geq 0$ and all $x \in X$ that $S^e(\varepsilon, x) + T^e(\varepsilon, x)$ is weak*-closed (in fact, the result works even for $\varepsilon_1 \neq \varepsilon_2$). When X is reflexive, or when S and T are strongly-representable, the regularity conditions given in Remark 4.16 turn out to be weaker than the one in [67]. Nevertheless, it is still an open question whether the condition $0 \in \text{sqri}\left(\text{pr}_X(\text{dom } \varphi_S) - \text{pr}_X(\text{dom } \varphi_T)\right)$ is in general sufficient for the weak*-closedness of the set $S^e(\varepsilon_1, x) + T^e(\varepsilon_2, x)$.

Chapter 5

Enlargements of positive sets

STEPHEN SIMONS proposed in [129] a more general setting for the study of monotone operators. He introduced the notion of a *positive set* with respect to a quadratic form q defined on a so-called *symmetrically self-dual Banach space (Banach SSD space)* (see also [130]) as an extension of the notion of a monotone set in Banach spaces. Several results from the theory of monotone operators have been successfully generalized to this more abstract framework. In his investigations SIMONS has mainly used some techniques based on the extension of the notion of *Fitzpatrick function* from the theory of monotone operators to a similar concept for positive sets. These investigations have been continued by the same author in [131], where notions and results recently introduced in the theory of monotone operators in general Banach spaces have known an appropriate generalization to positive sets in Banach SSD spaces. Let us mention here also the paper of MARTÍNEZ-LEGAZ [97], in which some further considerations and results are presented for maximally q-positive sets. Let us notice that the term *Simons space* is already used in the community when referring to the notion of *Banach SSD space* (see [23, 110]).

In analogy to the enlargement of a monotone operator, in this chapter we introduce and study the notion of enlargement of a positive set in (Banach) SSD spaces. The main results of this chapter are Theorem 5.1, where a one-to-one correspondence between a special family of enlargements of a maximally q-positive set and the family of representative functions associated to it is established, respectively Proposition 5.8, where it is shown that the extremal elements of the above mentioned family of representative functions associated to a maximally q-positive set are two functions recently introduced and studied by SIMONS in [131]. In this way we extend to (Banach) SSD spaces several results given for monotone and maximally monotone sets by BURACHIK AND SVAITER in [50, 51, 134]. The results from this chapter rely on [23].

5.1 Algebraic properties

Let us start by recalling the definition of an SSD space B along with some examples given in [131] and by giving a calculus rule for the quadratic form $q : B \to \mathbb{R}$ considered on it.

Definition 5.1 *(cf. [131, Definition 1.2]) (i) We say that $(B, \lfloor \cdot, \cdot \rfloor)$ is a symmetrically self-dual space (SSD space) if B is a nonzero vector space and $\lfloor \cdot, \cdot \rfloor : B \times B \to \mathbb{R}$ is a symmetric bilinear form. We consider the quadratic form $q : B \to \mathbb{R}$ defined by $q(b) = \frac{1}{2} \lfloor b, b \rfloor$ for all $b \in B$.*

(ii) A subset A of B is said to be q-positive if $A \neq \emptyset$ and $q(b - c) \geq 0$ for all

$b, c \in A$. *We say that A is* maximally *q-positive if A is q-positive and maximal (with respect to the inclusion) in the family of q-positive subsets of B.*

Remark 5.1 (a) In every SSD space B the following calculus rule is fulfilled:

$$q(\alpha b + \gamma c) = \alpha^2 q(b) + \gamma^2 q(c) + \alpha \gamma \lfloor b, c \rfloor \text{ for all } \alpha, \gamma \in \mathbb{R} \text{ and } b, c \in B. \quad (5.1)$$

(b) Let B be an SSD space and $A \subseteq B$ be a q-positive set. Then A is maximally q-positive if and only if for all $b \in B$ the implication below holds

$$q(b - c) \geq 0 \text{ for all } c \in A \Rightarrow b \in A.$$

Example 5.1 (cf. [131]) (a) If B is a Hilbert space with inner product $(b, c) \mapsto \langle b, c \rangle$ then B is an SSD space with $\lfloor b, c \rfloor = \langle b, c \rangle$ and $q(b) = \frac{1}{2} \|b\|^2$ and every non-empty subset of B is q-positive.

(b) If B is a Hilbert space with inner product $(b, c) \mapsto \langle b, c \rangle$ then B is an SSD space with $\lfloor b, c \rfloor = -\langle b, c \rangle$, $q(b) = -\frac{1}{2} \|b\|^2$ and the q-positive sets are the singletons.

(c) One can prove that $\mathbb{R}^3$ is an SSD space with $\lfloor (b_1, b_2, b_3), (c_1, c_2, c_3) \rfloor = b_1 c_2 + b_2 c_1 + b_3 c_3$ and $q(b_1, b_2, b_3) = b_1 b_2 + \frac{1}{2} b_3^2$. See [131] for a discussion regarding the q-positive sets in this setting.

(d) Consider X a nonzero Banach space and $B = X \times X^*$. For all $b = (x, x^*)$ and $c = (y, y^*) \in B$ we set $\lfloor b, c \rfloor = \langle y^*, x \rangle + \langle x^*, y \rangle$. Then B is an SSD space, $q(b) = \langle x^*, x \rangle$ and $q(b - c) = \langle x^* - y^*, x - y \rangle$. Hence for $A \subseteq B$ we have that A is q-positive exactly when A is a non-empty monotone subset of $X \times X^*$ in the usual sense and A is maximally q-positive exactly when A is a maximally monotone subset of $X \times X^*$ in the usual sense.

Let us consider in the following an arbitrary SSD space B and a function $f : B \to \overline{\mathbb{R}}$. We write $f^@$ for the conjugate of f with respect to the pairing $\lfloor \cdot, \cdot \rfloor$, that is $f^@(c) = \sup_{b \in B}\{\lfloor c, b \rfloor - f(b)\}$. We write $\mathcal{P}(f) = \{b \in B : f(b) = q(b)\}$. If f is proper and convex, $f \geq q$ on B and $\mathcal{P}(f) \neq \emptyset$, then $\mathcal{P}(f)$ is a q-positive subset of B (see [131, Lemma 1.9]). Conditions under which $\mathcal{P}(f)$ is maximally q-positive are given in [131, Theorem 2.9].

We introduce in the following the concept of enlargement of a positive set and study some of its algebraic properties.

Definition 5.2 *Let B be an SSD space. Given A a q-positive subset of B, we say that the multifunction $E : \mathbb{R}_+ \rightrightarrows B$ is an* enlargement *of A if*

$$A \subseteq E(\varepsilon) \text{ for all } \varepsilon \geq 0.$$

Example 5.2 Let B be an SSD space and A a q-positive subset of B. The multifunction $E^A : \mathbb{R}_+ \rightrightarrows B$ defined by

$$E^A(\varepsilon) = \{b \in B : q(b - c) \geq -\varepsilon \text{ for all } c \in A\}$$

is an enlargement of A. Let us notice that A is maximally q-positive if and only if $A = E^A(0)$. Moreover, in the framework of Example 5.1(d), for the graph of E^A we have $G(E^A) = \{(\varepsilon, x, x^*) : \langle x^* - y^*, x - y \rangle \geq -\varepsilon \text{ for all } (y, y^*) \in A\}$, hence in this case $G(E^A) = G(A^e)$ (see Section 4.2 for the definition of A^e).

The following definition extends to SSD spaces a notion given in [49,50] (see also [51, Definition 2.3] and Definition 4.3).

Definition 5.3 *We say that the multifunction $E : \mathbb{R}_+ \rightrightarrows B$ satisfies the* transportation formula *if for every $\varepsilon_1, \varepsilon_2 \geq 0$, $b^1 \in E(\varepsilon_1)$, $b^2 \in E(\varepsilon_2)$ and every $\alpha_1, \alpha_2 \geq 0$, $\alpha_1 + \alpha_2 = 1$ we have $\varepsilon := \alpha_1 \varepsilon_1 + \alpha_2 \varepsilon_2 + \alpha_1 \alpha_2 q(b^1 - b^2) \geq 0$ and $\alpha_1 b^1 + \alpha_2 b^2 \in E(\varepsilon)$.*

Proposition 5.1 *Let B be an SSD space and $A \subseteq B$ be a maximally q-positive set. Then E^A satisfies the transportation formula.*

Proof. Take $\varepsilon_1, \varepsilon_2 \geq 0$, $b^1 \in E^A(\varepsilon_1)$, $b^2 \in E^A(\varepsilon_2)$ and $\alpha_1, \alpha_2 \geq 0$, $\alpha_1 + \alpha_2 = 1$. We have to show that

$$q(\alpha_1 b^1 + \alpha_2 b^2 - c) \geq -\alpha_1 \varepsilon_1 - \alpha_2 \varepsilon_2 - \alpha_1 \alpha_2 q(b^1 - b^2) \text{ for all } c \in A \qquad (5.\ 2)$$

and

$$\alpha_1 \varepsilon_1 + \alpha_2 \varepsilon_2 + \alpha_1 \alpha_2 q(b^1 - b^2) \geq 0. \qquad (5.\ 3)$$

Let $c \in A$ be arbitrary. By using the inequalities $q(b^1 - c) \geq -\varepsilon_1$ and $q(b^2 - c) \geq -\varepsilon_2$ and the calculus rule (5. 1) we obtain

$$q(\alpha_1 b^1 + \alpha_2 b^2 - c) = q\big(\alpha_1(b^1 - c) + \alpha_2(b^2 - c)\big) = \alpha_1^2 q(b^1 - c) + \alpha_2^2 q(b^2 - c)$$

$$+\alpha_1 \alpha_2 \lfloor b^1 - c, b^2 - c \rfloor = \alpha_1^2 q(b^1 - c) + \alpha_2^2 q(b^2 - c) + \alpha_1 \alpha_2 \big(q(b^1 - c) + q(b^2 - c) - q(b^1 - b^2)\big)$$

$$= \alpha_1 q(b^1 - c) + \alpha_2 q(b^2 - c) - \alpha_1 \alpha_2 q(b^1 - b^2) \geq -\alpha_1 \varepsilon_1 - \alpha_2 \varepsilon_2 - \alpha_1 \alpha_2 q(b^1 - b^2).$$

Let us suppose that $\alpha_1 \varepsilon_1 + \alpha_2 \varepsilon_2 + \alpha_1 \alpha_2 q(b^1 - b^2) < 0$. From (5. 2) we obtain

$$q(\alpha_1 b^1 + \alpha_2 b^2 - c) > 0 \text{ for all } c \in A. \qquad (5.\ 4)$$

Since A is maximally q-positive we get $\alpha_1 b^1 + \alpha_2 b^2 \in A$ (cf. Remark 5.1(b)). By choosing $c := \alpha_1 b^1 + \alpha_2 b^2 \in A$ in (5. 4) we get $0 > 0$, which is a contradiction. Hence (5. 3) is also fulfilled and the proof is complete. $\qquad \square$

The following result establishes a connection between the transportation formula and convexity (see also [134, Lemma 3.2]).

Proposition 5.2 *Let B be an SSD space, $E : \mathbb{R}_+ \rightrightarrows B$ a multifunction and define the function $\Psi : \mathbb{R} \times B \to \mathbb{R} \times B$, $\Psi(\varepsilon, b) = (\varepsilon + q(b), b)$ for all $(\varepsilon, b) \in \mathbb{R} \times B$. The following statements are equivalent:*

(i) E satisfies the transportation formula;

(ii) E satisfies the generalized transportation formula (or the n-point transportation formula), that is for all $n \geq 1$, $\varepsilon_i \geq 0$, $b^i \in E(\varepsilon_i)$ and $\alpha_i \geq 0$, $i = 1, ..., n$, with $\sum_{i=1}^n \alpha_i = 1$ we have $\varepsilon := \sum_{i=1}^n \alpha_i \varepsilon_i + \sum_{i=1}^n \alpha_i q\big(b^i - \sum_{j=1}^n \alpha_j b^j\big) \geq 0$ and $\sum_{i=1}^n \alpha_i b^i \in E(\varepsilon)$;

(iii) $\Psi(G(E))$ is a convex subset of $\mathbb{R} \times B$.

Proof. We notice first that Ψ is a bijective function with inverse $\Psi^{-1} : \mathbb{R} \times B \to \mathbb{R} \times B$, $\Psi^{-1}(\varepsilon, b) = (\varepsilon - q(b), b)$ for all $(\varepsilon, b) \in \mathbb{R} \times B$.

(ii)$\Rightarrow$(i) Take $\varepsilon_1, \varepsilon_2 \geq 0$, $b^1 \in E(\varepsilon_1)$, $b^2 \in E(\varepsilon_2)$ and $\alpha_1, \alpha_2 \geq 0$, $\alpha_1 + \alpha_2 = 1$. Then for $\varepsilon := \alpha_1 \varepsilon_1 + \alpha_2 \varepsilon_2 + \alpha_1 q\big(b^1 - (\alpha_1 b^1 + \alpha_2 b^2)\big) + \alpha_2 q\big(b^2 - (\alpha_1 b^1 + \alpha_2 b^2)\big) \geq 0$ we get $\alpha_1 b^1 + \alpha_2 b^2 \in E(\varepsilon)$. Since

$$\varepsilon = \alpha_1 \varepsilon_1 + \alpha_2 \varepsilon_2 + \alpha_1 \alpha_2^2 q(b^1 - b^2) + \alpha_2 \alpha_1^2 q(b^1 - b^2) = \alpha_1 \varepsilon_1 + \alpha_2 \varepsilon_2 + \alpha_1 \alpha_2 q(b^1 - b^2),$$

this implies that E satisfies the transportation formula.

(i)$\Rightarrow$(iii) Let be $(\mu_1, b^1), (\mu_2, b^2) \in \Psi(G(E))$ and $\alpha_1, \alpha_2 \geq 0$ with $\alpha_1 + \alpha_2 = 1$. Then there exist $\varepsilon_1, \varepsilon_2 \geq 0$ such that $\mu_1 = \varepsilon_1 + q(b^1)$, $b^1 \in E(\varepsilon_1)$ and $\mu_2 = \varepsilon_2 + q(b^2)$, $b^2 \in E(\varepsilon_2)$. By (i) we have that $\varepsilon := \alpha_1 \varepsilon_1 + \alpha_2 \varepsilon_2 + \alpha_1 \alpha_2 q(b^1 - b^2) \geq 0$ and $\alpha_1 b^1 + \alpha_2 b^2 \in E(\varepsilon)$. Using (5. 1) we further get that

$$\varepsilon + q(\alpha_1 b^1 + \alpha_2 b^2) = \alpha_1 \varepsilon_1 + \alpha_2 \varepsilon_2 + \alpha_1 \alpha_2 \big(q(b^1) + q(b^2) - \lfloor b^1, b^2 \rfloor\big) + \alpha_1^2 q(b^1) + \alpha_2^2 q(b^2) +$$

$$\alpha_1 \alpha_2 \lfloor b^1, b^2 \rfloor = \alpha_1 \varepsilon_1 + \alpha_2 \varepsilon_2 + \alpha_1 q(b^1) + \alpha_2 q(b^2) = \alpha_1 \mu_1 + \alpha_2 \mu_2.$$

Thus

$$\alpha_1 (\mu_1, b^1) + \alpha_2 (\mu_2, b^2) = \big(\varepsilon + q(\alpha_1 b^1 + \alpha_2 b^2), \alpha_1 b^1 + \alpha_2 b^2 \big)$$

$$= \Psi(\varepsilon, \alpha_1 b^1 + \alpha_2 b^2) \in \Psi(G(E))$$

and this provides the convexity of $\Psi(G(E))$.

(iii)$\Rightarrow$(ii) Let be $n \geq 1$, $\varepsilon_i \geq 0$, $b^i \in E(\varepsilon_i)$ and $\alpha_i \geq 0$, $i = 1, ..., n$, with $\sum_{i=1}^{n} \alpha_i = 1$. This means that $(\varepsilon_i + q(b^i), b^i) \in \Psi(G(E))$ for $i = 1, ..., n$. Let us make the following notations: $b := \sum_{i=1}^{n} \alpha_i b^i$ and $\varepsilon := \sum_{i=1}^{n} \alpha_i \big(\varepsilon_i + q(b^i) \big) - q(b)$. By using the convexity of $\Psi(G(E))$ one has $(\varepsilon + q(b), b) = \sum_{i=1}^{n} \alpha_i \big(\varepsilon_i + q(b^i), b^i \big) \in \Psi(G(E))$, which implies that $\Psi^{-1}(\varepsilon + q(b), b) = (\varepsilon, b) \in G(E)$. From here if follows that $\varepsilon = \sum_{i=1}^{n} \alpha_i \varepsilon_i + \sum_{i=1}^{n} \alpha_i q(b^i) - q(b) \geq 0$ and $b = \sum_{i=1}^{n} \alpha_i b^i \in E(\varepsilon)$. To finish the proof we have only to show that $\sum_{i=1}^{n} \alpha_i q(b^i - b) = \sum_{i=1}^{n} \alpha_i q(b^i) - q(b)$. Indeed,

$$\sum_{i=1}^{n} \alpha_i q(b^i - b) = \sum_{i=1}^{n} \alpha_i q(b^i) + \sum_{i=1}^{n} \alpha_i q(b) - \sum_{i=1}^{n} \alpha_i \lfloor b^i, b \rfloor = \sum_{i=1}^{n} \alpha_i q(b^i) + q(b) - \lfloor b, b \rfloor$$

$$= \sum_{i=1}^{n} \alpha_i q(b^i) + q(b) - 2q(b) = \sum_{i=1}^{n} \alpha_i q(b^i) - q(b)$$

and, consequently, the generalized transportation formula holds. $\qquad\square$

Like in [134, Definition 3.3] (see also [51, Definition 2.5]) one can introduce a family of enlargements associated to a positive set (see also Definition 4.3).

Definition 5.4 *Let B be an SSD space and $A \subseteq B$ be a q-positive set. We define $\mathbb{E}(A)$ as being the family of multifunctions $E : \mathbb{R}_+ \rightrightarrows B$ satisfying the following properties:*

(r1) E is an enlargement of A, that is

$$A \subseteq E(\varepsilon) \text{ for all } \varepsilon \geq 0;$$

(r2) E is nondecreasing, that is

$$0 \leq \varepsilon_1 \leq \varepsilon_2 \Rightarrow E(\varepsilon_1) \subseteq E(\varepsilon_2);$$

(r3) E satisfies the transportation formula.

If A is maximally q-positive then E^A satisfies the properties (r1)-(r3) (cf. Example 5.2 and Proposition 5.1, while (r2) is obviously satisfied), hence in this case the family $\mathbb{E}(A)$ is non-empty. Let us define the multifunction $E_A : \mathbb{R}_+ \rightrightarrows B$, $E_A(\varepsilon) := \bigcap_{E \in \mathbb{E}(A)} E(\varepsilon)$ for all $\varepsilon \geq 0$.

Proposition 5.3 *Let B be an SSD space and $A \subseteq B$ be a maximally q-positive set. Then:*

(i) $E_A, E^A \in \mathbb{E}(A)$;

(ii) E_A and E^A are, respectively, the smallest and the biggest elements in $\mathbb{E}(A)$ with respect to the partial ordering inclusion relation of the graphs, that is $G(E_A) \subseteq G(E) \subseteq G(E^A)$ for all $E \in \mathbb{E}(A)$.

Proof. (i) That $E^A \in \mathbb{E}(A)$ was pointed out above. The statement $E_A \in \mathbb{E}(A)$ follows immediately, if we take into consideration the definition of E_A.

(ii) E_A is obviously the smallest element in $\mathbb{E}(A)$. We prove in the following that E^A is the biggest element in $\mathbb{E}(A)$. Suppose that E^A is not the biggest element in $\mathbb{E}(A)$, namely that there exist $E \in \mathbb{E}(A)$ and $(\varepsilon, b) \in G(E) \setminus G(E^A)$. Since $(\varepsilon, b) \notin G(E^A)$, there exists $c \in A$ such that $q(b - c) < -\varepsilon$. Let $\lambda \in (0, 1)$ be fixed. As E satisfies (r1), we have $c \in A \subseteq E(0)$, that is $(0, c) \in G(E)$. As $(\varepsilon, b), (0, c) \in G(E)$, $\lambda \in (0, 1)$ and E satisfies the transportation formula, we obtain $\lambda \varepsilon + \lambda(1 - \lambda)q(b - c) \geq 0$, hence $\varepsilon + (1 - \lambda)q(b - c) \geq 0$. Since this inequality must hold for arbitrary $\lambda \in (0, 1)$, we get $\varepsilon + q(b - c) \geq 0$, which is a contradiction. $\square$

Lemma 5.1 *Let B be an SSD space, $A \subseteq B$ a maximally q-positive set and $E \in \mathbb{E}(A)$. Then*

$$E(0) = \bigcap_{\varepsilon > 0} E(\varepsilon) = A.$$

Proof. By using the properties (r1) and (r2), Proposition 5.3, the definition of E^A and Example 5.2 we get

$$A \subseteq E(0) \subseteq \bigcap_{\varepsilon > 0} E(\varepsilon) \subseteq \bigcap_{\varepsilon > 0} E^A(\varepsilon) = E^A(0) = A$$

and the conclusion follows. $\square$

5.2   Topological properties

We start by recalling the definition of a Banach SSD space, a concept introduced by STEPHEN SIMONS, and further we study some topological properties of enlargements of positive sets in this framework.

Definition 5.5 *We say that B is a Banach SSD space if B is an SSD space and $\|\cdot\|$ is a norm on B with respect to which B is a Banach space with norm-dual B^*,*

$$\frac{1}{2}\|\cdot\|^2 + q \geq 0 \text{ on } B \tag{5.5}$$

and there exists $\iota : B \to B^*$ linear and continuous such that

$$\langle \iota(c), b \rangle = \lfloor b, c \rfloor \text{ for all } b, c \in B. \tag{5.6}$$

Remark 5.2 (i) From (5.6) we obtain $|\lfloor b, c \rfloor| \leq \|\iota\|\|b\|\|c\|$. Hence for $(b, c), (\bar{b}, \bar{c}) \in B \times B$ it holds

$$|\lfloor b, c \rfloor - \lfloor \bar{b}, \bar{c} \rfloor| = |\lfloor b - \bar{b}, c - \bar{c} \rfloor + \lfloor \bar{b}, c - \bar{c} \rfloor + \lfloor b - \bar{b}, \bar{c} \rfloor|$$

$$\leq \|\iota\|(\|b - \bar{b}\|\|c - \bar{c}\| + \|\bar{b}\|\|c - \bar{c}\| + \|b - \bar{b}\|\|\bar{c}\|).$$

The function $(b, c) \mapsto \lfloor b, c \rfloor$ is, consequently, continuous and from here one gets immediately the continuity of q, $\lfloor \cdot, c \rfloor$ and $\lfloor b, \cdot \rfloor$ for all $b, c \in B$.

(ii) For a function $f : B \to \overline{\mathbb{R}}$ we have $f^@(c) = \sup_{b \in B}\{\langle \iota(c), b \rangle - f(b)\} = f^*(\iota(c))$, that is $f^@ = f^* \circ \iota$ on B.

Example 5.3 (a) The SSD spaces considered in Example 5.1(a)-(c) are Banach SSD spaces (see [131, Remark 2.2]).

(b) Consider again the framework of Example 5.1(d), that is X is a nonzero Banach space and $B = X \times X^*$. The canonical embedding of X into X^{**} is defined by $\hat{\ } : X \to X^{**}$, $\langle \hat{x}, x^* \rangle := \langle x^*, x \rangle$ for all $x \in X$ and all $x^* \in X^*$. The dual of B (with respect to the norm topology) is $X^* \times X^{**}$ under the pairing

$$\langle c^*, b \rangle = \langle y^*, x \rangle + \langle y^{**}, x^* \rangle \ \forall b = (x, x^*) \in B \ \forall c^* = (y^*, y^{**}) \in B^*.$$

Thus $X \times X^*$ is a Banach SSD space, where $\iota : X \times X^* \to X^* \times X^{**}$, $\iota(x, x^*) = (x^*, \hat{x})$ for all $(x, x^*) \in X \times X^*$.

For $E : \mathbb{R}_+ \rightrightarrows B$ we define $\overline{E} : \mathbb{R}_+ \rightrightarrows B$ by $\overline{E}(\varepsilon) := \{b \in B : (\varepsilon, b) \in \text{cl}\,(G(E))\}$. The multifunction E is said to be closed if $E = \overline{E}$. One can see that E is closed if and only if $G(E)$ is closed. For $A \subseteq B$, consider also the subfamily $\mathbb{E}_c(A) = \{E \in \mathbb{E}(A) : E \text{ is closed}\}$.

Proposition 5.4 *Let B be a Banach SSD space and $A \subseteq B$ be a maximally q-positive set. The following statements are true:*

(i) *If $E \in \mathbb{E}(A)$ then $\overline{E} \in \mathbb{E}_c(A)$.*

(ii) *If $E \in \mathbb{E}_c(A)$ then $E(\varepsilon)$ is closed, for all $\varepsilon \geq 0$.*

(iii) *$\overline{E_A}$ and E^A are, respectively, the smallest and the biggest elements in $\mathbb{E}_c(A)$, with respect to the partial ordering inclusion relation of the graphs, that is $G(\overline{E_A}) \subseteq G(E) \subseteq G(E^A)$ for all $E \in \mathbb{E}_c(A)$.*

Proof. (i) Let be $E \in \mathbb{E}(A)$. One can notice that the continuity of the function q implies that if E satisfies the transportation formula, then $\overline{E}$ satisfies this formula, too. Further, if E is nondecreasing, then $\overline{E}$ is also nondecreasing. Hence the first assertion follows.

(ii) The second statement of the proposition is a consequence of the fact that E is closed if and only if $G(E)$ is closed.

(iii) Employing once more the continuity of the function q we get that $G(E^A)$ is closed. Combining Proposition 5.3 and Proposition 5.4(i) we obtain $\overline{E_A}, E^A \in \mathbb{E}_c(A)$. The proof of the minimality, respectively, maximality of these elements presents no difficulty. $\qquad\square$

In the following we establish a one-to-one correspondence between $\mathbb{E}_c(A)$ and a family of convex functions associated to A. This is done be extending to Banach SSD spaces the techniques used by BURACHIK AND SVAITER in [51, Section 3].

Consider B a Banach SSD space. To $A \subseteq B \times \mathbb{R}$ we associate the so-called *lower envelope of A* (cf. [4]), defined as $\text{le}_A : B \to \overline{\mathbb{R}}$, $\text{le}_A(b) = \inf\{r \in \mathbb{R} : (b, r) \in A\}$. Obviously, $A \subseteq \text{epi}(\text{le}_A)$. If, additionally, A is closed and has an *epigraphical structure*, that is $(b, r_1) \in A \Rightarrow (b, r_2) \in A$ for all $r_2 \in [r_1, +\infty)$, then $A = \text{epi}(\text{le}_A)$.

Let us consider now a multifunction $E : \mathbb{R}_+ \rightrightarrows B$ and define $\lambda_E : B \to \overline{\mathbb{R}}$, $\lambda_E(b) = \inf\{\varepsilon \geq 0 : b \in E(\varepsilon)\}$. It is easy to observe that $\lambda_E(b) = \inf\{r \in \mathbb{R} : (b, r) \in G(E^{-1})\}$, where $E^{-1} : B \rightrightarrows \mathbb{R}_+$ is the *inverse of the multifunction E*. One has $G(E^{-1}) = \{(b, \varepsilon) : (\varepsilon, b) \in G(E)\}$. Hence λ_E is the lower envelope of $G(E^{-1})$. We have $G(E^{-1}) \subseteq \text{epi}(\lambda_E)$. If E is closed and nondecreasing, then $G(E^{-1})$ is closed and has an epigraphical structure, so in this case $G(E^{-1}) = \text{epi}(\lambda_E)$. As in [51, Proposition 3.1] we obtain the following result.

Proposition 5.5 *Let B be a Banach SSD space and $E : \mathbb{R}_+ \rightrightarrows B$ be a multifunction which is closed and nondecreasing. Then*

(i) *$G(E^{-1}) = \text{epi}(\lambda_E)$;*

(ii) *λ_E is lower semicontinuous;*

(iii) $\lambda_E \geq 0$;

(iv) $E(\varepsilon) = \{b \in B : \lambda_E(b) \leq \varepsilon\}$ *for all* $\varepsilon \geq 0$.

Moreover, λ_E *is the only function from* B *to* $\overline{\mathbb{R}}$ *satisfying (iii) and (iv).*

Given $E : \mathbb{R}_+ \rightrightarrows B$, we define the function $\Lambda_E : B \to \overline{\mathbb{R}}$, $\Lambda_E := \lambda_E + q$. Let us notice that Λ_E is the lower envelope of $\Psi(G(E^{-1}))$ (the function Ψ was defined in Proposition 5.2) and $\mathrm{epi}(\Lambda_E) = \Psi(\mathrm{epi}(\lambda_E))$. From these observations, Proposition 5.5(i) and Proposition 5.2 we obtain the following result.

Corollary 5.1 *Let B be Banach SSD space and $E : \mathbb{R}_+ \rightrightarrows B$ a closed and nondecreasing enlargement of the maximally q-positive set $A \subseteq B$. Then $E \in \mathbb{E}(A)$ if and only if Λ_E is convex.*

Proposition 5.6 *Let B be a Banach SSD space, $A \subseteq B$ a maximally q-positive set and $E \in \mathbb{E}_c(A)$. Then Λ_E is convex, lower semicontinuous, $\Lambda_E \geq q$ on B and $A \subseteq \mathcal{P}(\Lambda_E)$.*

Proof. The first three assertions follow from Corollary 5.1 and Proposition 5.5(ii) and (iii). Take an arbitrary $b \in A$. Since E is an enlargement of A we get $b \in E(0)$, hence $\lambda_E(b) = 0$ and the conclusion follows. $\qquad\square$

To every maximally q-positive set we introduce the following family of convex functions (compare with Definition 4.1).

Definition 5.6 *Let B be a Banach SSD space and $A \subseteq B$ be a maximally q-positive set. We define $\mathcal{H}(A)$ as the family of convex and lower semicontinuous functions $h : B \to \overline{\mathbb{R}}$ such that*

$$h \geq q \text{ on } B \text{ and } A \subseteq \mathcal{P}(h).$$

Remark 5.3 *Combining Proposition 5.6 and Proposition 5.5(i) we obtain that the map $E \mapsto \Lambda_E$ is one-to-one from $\mathbb{E}_c(A)$ to $\mathcal{H}(A)$.*

For $h \in \mathcal{H}(A)$ we define the multifunction $A_h : \mathbb{R}_+ \rightrightarrows B$,

$$A_h(\varepsilon) := \{b \in B : h(b) \leq \varepsilon + q(b)\} \text{ for all } \varepsilon \geq 0.$$

Proposition 5.7 *Let B be a Banach SSD space and $A \subseteq B$ be a maximally q-positive set. If $h \in \mathcal{H}(A)$, then $A_h \in \mathbb{E}_c(A)$ and $\Lambda_{A_h} = h$.*

Proof. Take an arbitrary $h \in \mathcal{H}(A)$. The properties of the function h imply that A_h is a closed enlargement of A. Obviously A_h is nondecreasing. Trivially, $A_h(\varepsilon) = \{b \in B : l(b) \leq \varepsilon\}$, where $l : B \to \overline{\mathbb{R}}$, $l := h - q$. By Proposition 5.5 we get $\lambda_{A_h} = l$, implying $\Lambda_{A_h} = h$. The convexity of h and Corollary 5.1 guarantee that $A_h \in \mathbb{E}_c(A)$. $\qquad\square$

As a consequence of the above results we obtain a bijection between the family of closed enlargements (which satisfy condition (r1)-(r3) from Definition 5.4) associated to a maximally q-positive set and the family of convex functions introduced in Definition 5.6.

Theorem 5.1 *Let B be a Banach SSD space and $A \subseteq B$ be a maximally q-positive set. The map*

$$\mathbb{E}_c(A) \to \mathcal{H}(A),$$

$$E \mapsto \Lambda_E$$

is a bijection, with inverse given by

$$\mathcal{H}(A) \to \mathbb{E}_c(A),$$

$$h \mapsto A_h.$$

Moreover, $A_{\Lambda_E} = E$ for all $E \in \mathbb{E}_c(A)$ and $\Lambda_{A_h} = h$ for all $h \in \mathcal{H}(A)$.

The following corollary shows that for a maximally q-positive set A, the elements of $\mathcal{H}(A)$ are closely related to A.

Corollary 5.2 *Let B be a Banach SSD space and $A \subseteq B$ be a maximally q-positive set. Take $h \in \mathcal{H}(A)$. Then $A = \mathcal{P}(h)$.*

Proof. We have $A \subseteq \mathcal{P}(h)$ by the definition of $\mathcal{H}(A)$. Take an arbitrary $b \in \mathcal{P}(h)$ and define $E := A_h$. Then $b \in E(0)$. Applying Theorem 5.1 we get $E \in \mathbb{E}_c(A)$. Further, by Lemma 5.1 we have $E(0) = A$, hence $b \in A$ and the proof is complete. $\square$

Remark 5.4 In what follows, we call an arbitrary element h of $\mathcal{H}(A)$ a *representative function* of A. The word "representative" is justified by Corollary 5.2. Since for A a q-positive set, we have $A \neq \emptyset$ (see Definition 5.1(ii)), every representative function of A is proper.

Corollary 5.3 *Let B be a Banach SSD space and $A \subseteq B$ be a maximally q-positive set. Take $E \in \mathbb{E}_c(A)$ and $b^1 \in E(\varepsilon_1), b^2 \in E(\varepsilon_2)$, where $\varepsilon_1, \varepsilon_2 \geq 0$ are arbitrary. Then*

$$q(b^1 - b^2) \geq -(\sqrt{\varepsilon_1} + \sqrt{\varepsilon_2})^2.$$

Proof. By Theorem 5.1, there exists a representative function $h \in \mathcal{H}(A)$ such that $E = A_h$. By using the definition of A_h and by applying [131, Lemma 1.6] we obtain

$$-q(b^1 - b^2) \leq \left[\sqrt{(h-q)(b^1)} + \sqrt{(h-q)(b^2)}\right]^2 \leq (\sqrt{\varepsilon_1} + \sqrt{\varepsilon_2})^2$$

and the proof is complete. $\square$

Remark 5.5 In case B is taken as in Example 5.3(b) and $E = A^e$ (see Example 5.2), the above lower bound is established in [50, Corollary 3.12]. Here we generalize this result to Banach SSD spaces and to an arbitrary $E \in \mathbb{E}_c(A)$.

In the following we investigate the properties of the functions Λ_{E^A} and $\Lambda_{\overline{E_A}}$ and rediscover in this way the functions introduced and studied by S. Simons in [129–131] (see Proposition 5.8(iii) from below).

Corollary 5.4 *Let B be a Banach SSD space and $A \subseteq B$ be a maximally q-positive set.*

(i) *The functions Λ_{E^A} and $\Lambda_{\overline{E_A}} \in \mathcal{H}(A)$ and are, respectively, the minimum and the maximum of this family, that is*

$$\Lambda_{E^A} \leq h \leq \Lambda_{\overline{E_A}} \text{ for all } h \in \mathcal{H}(A). \tag{5.7}$$

(ii) *Conversely, if $h : B \to \overline{\mathbb{R}}$ is a convex and lower semicontinuous function such that*

$$\Lambda_{E^A} \leq h \leq \Lambda_{\overline{E_A}}, \tag{5.8}$$

then $h \in \mathcal{H}(A)$.

(iii) It holds $\mathcal{H}(A) = \{h : B \to \overline{\mathbb{R}} \mid h \text{ convex, lower semicontinuous and } \Lambda_{E^A} \le h \le \Lambda_{\overline{E_A}}\}$.

Proof. (i) This follows immediately from Theorem 5.1 and Proposition 5.4.

(ii) If $h : B \to \overline{\mathbb{R}}$ is a convex and lower semicontinuous function satisfying (5. 8), then (since $\Lambda_{E^A} \in \mathcal{H}(A)$)

$$h \ge \Lambda_{E^A} \ge q \text{ on } B. \tag{5.9}$$

Further, for $b \in A$ we obtain (employing that $\Lambda_{\overline{E_A}} \in \mathcal{H}(A)$) that $h(b) \le \Lambda_{\overline{E_A}}(b) = q(b)$. In view of (5. 9) it follows that $b \in \mathcal{P}(h)$, hence $h \in \mathcal{H}(A)$.

(iii) This characterization of $\mathcal{H}(A)$ is a direct consequence of (i) and (ii). $\square$

Definition 5.7 *(cf. [131]) Let B be a Banach SSD space and $A \subseteq B$ be a q-positive set. We define the functions $\Theta_A : B^* \to \overline{\mathbb{R}}$, $\Phi_A : B \to \overline{\mathbb{R}}$ and ${}^*\Theta_A : B \to \overline{\mathbb{R}}$ by*

$$\Theta_A(b^*) := \sup_{a \in A}\{\langle b^*, a\rangle - q(a)\} \text{ for all } b^* \in B^*,$$

$$\Phi_A := \Theta_A \circ \iota$$

and, respectively,

$${}^*\Theta_A(c) := \sup_{b^* \in B^*} \{\langle b^*, c\rangle - \Theta_A(b^*)\} \text{ for all } c \in B.$$

By $a \vee b$ we denote the maximum value of $a, b \in \overline{\mathbb{R}}$. The following properties of the functions defined above appear in [131, Lemma 2.13 and Theorem 2.16]. The property (vii) is a direct consequence of (i)-(vi).

Lemma 5.2 *Let B be a Banach SSD space and $A \subseteq B$ be a q-positive set. Then*

(i) For all $b \in B$, $\Phi_A(b) = \sup_{a \in A}\{\lfloor b, a\rfloor - q(a)\} = q(b) - \inf_{c \in A} q(b - c)$.

(ii) Φ_A is proper, convex, lower semicontinuous and $A \subseteq \mathcal{P}(\Phi_A)$.

(iii) $({}^*\Theta_A)^* = \Theta_A$ and $({}^*\Theta_A)^@ = \Phi_A$.

(iv) ${}^*\Theta_A$ is proper, convex, lower semicontinuous, ${}^*\Theta_A \ge \Phi_A^@ \ge \Phi_A \vee q$ on B and

$${}^*\Theta_A = \Phi_A^@ = q \text{ on } A.$$

(v) ${}^*\Theta_A = \sup\{h : B \to \overline{\mathbb{R}} \mid h \text{ proper, convex, lower semicontinuous, } h \le q \text{ on } A\}$.

If, additionally, A is maximally q-positive, then

(vi) ${}^*\Theta_A \ge \Phi_A^@ \ge \Phi_A \ge q$ on B and $A = \mathcal{P}({}^*\Theta_A) = \mathcal{P}(\Phi_A^@) = \mathcal{P}(\Phi_A)$.

(vii) ${}^*\Theta_A, \Phi_A^@, \Phi_A \in \mathcal{H}(A)$.

Next we give other characterizations of the function ${}^*\Theta_A$ and establish the connection between $\Lambda_{E^A}, \Lambda_{\overline{E_A}}$ and $\Phi_A, {}^*\Theta_A$, respectively.

Proposition 5.8 *Let B be a Banach SSD space and $A \subseteq B$ be a q-positive set. Then*

(i) ${}^*\Theta_A = \sup\{h : B \to \overline{\mathbb{R}} \mid h \text{ proper, convex, lower semicontinuous, } h \ge q \text{ on } B \text{ and } A \subseteq \mathcal{P}(h)\}$.

(ii) ${}^*\Theta_A = \operatorname{cl}\operatorname{co}(q + \delta_A)$.

If, additionally, A is maximally q-positive, then

(iii) $\Lambda_{E^A} = \Phi_A$ and $\Lambda_{\overline{E_A}} = {}^*\Theta_A$.

(iv) If $h : B \to \overline{\mathbb{R}}$ is a function such that $h \in \mathcal{H}(A)$, then $h^@ \in \mathcal{H}(A)$.

Proof. (i) We have

$$\{h : B \to \overline{\mathbb{R}} \mid h \text{ proper, convex, lower semicontinuous}, h \geq q \text{ on } B \text{ and } A \subseteq \mathcal{P}(h)\}$$

$$\subseteq \{h : B \to \overline{\mathbb{R}} \mid h \text{ proper, convex, lower semicontinuous}, h \leq q \text{ on } A\},$$

hence from Lemma 5.2(v) we get

$$\sup\{h : B \to \overline{\mathbb{R}} \mid h \text{ proper, convex, lower semicontinuous},$$

$$h \geq q \text{ on } B \text{ and } A \subseteq \mathcal{P}(h)\} \leq {}^*\Theta_A.$$

On the other hand, by Lemma 5.2(iv), ${}^*\Theta_A$ is proper, convex and lower semicontinuous and it fulfills ${}^*\Theta_A \geq q$ on B and $A \subseteq \mathcal{P}({}^*\Theta_A)$. Thus the equality follows.

(ii) Since ${}^*\Theta_A \leq q$ on A we have ${}^*\Theta_A \leq q + \delta_A$ on B, hence

$$ {}^*\Theta_A \leq \text{cl}\,\text{co}(q + \delta_A) \leq q + \delta_A. \tag{5. 10}$$

The above inequality shows that $\text{cl}\,\text{co}(q + \delta_A)$ is a proper, convex, lower semicontinuous function such that $\text{cl}\,\text{co}(q + \delta_A) \leq q$ on A. Applying Lemma 5.2(v) we obtain ${}^*\Theta_A \geq \text{cl}\,\text{co}(q + \delta_A)$, which combined with (5. 10) delivers the desired result.

(iii) From Lemma 5.2(i) and the definition of E^A we obtain

$$b \in E^A(\varepsilon) \Leftrightarrow q(b - c) \geq -\varepsilon \text{ for all } c \in A \Leftrightarrow \inf_{c \in A} q(b - c) \geq -\varepsilon$$

$$\Leftrightarrow q(b) - \Phi_A(b) \geq -\varepsilon \Leftrightarrow \Phi_A(b) \leq \varepsilon + q(b).$$

This is nothing else than $E^A = A_{\Phi_A}$. Theorem 5.1 implies that $\Lambda_{E^A} = \Lambda_{A_{\Phi_A}} = \Phi_A$.

The equality $\Lambda_{\overline{E_A}} = {}^*\Theta_A$ follows from (i) and Corollary 5.4.

(iv) From (iii), Corollary 5.4 and [131, Theorem 2.15 (b)] we get $h^@ \geq q$ on B and $\mathcal{P}(h) = \mathcal{P}(h^@) = A$. The function $h^@$ is proper and convex, while its lower semicontinuity follows from the definition of $h^@$ and Remark 5.2, hence $h^@ \in \mathcal{H}(A)$. □

Remark 5.6 Proposition 5.8(iv) is a generalization of [51, Theorem 5.3] to Banach SSD spaces.

Remark 5.7 In general, the functions ${}^*\Theta_A$ and $\Phi_A^@$ are not identical. A striking example in this sense was given by C. ZĂLINESCU (see [131, Remark 2.14]) for B a Banach space and $\lfloor \cdot, \cdot \rfloor = 0$ on $B \times B$. An alternative example, considered by M.D. VOISEI AND C. ZĂLINESCU in another context, is given below (see Example 5.4).

Before we present this example, we need the following remark.

Remark 5.8 Consider again the particular setting of Example 5.1(d) and Example 5.3(b), namely when $B = X \times X^*$, where X is a nonzero Banach space. Let A be a non-empty monotone subset of $X \times X^*$. In this case $q(x, x^*) = \langle x^*, x \rangle$ for all $(x, x^*) \in X \times X^*$ and the function $\Theta_A : X^* \times X^{**} \to \overline{\mathbb{R}}$ is defined by

$$\Theta_A(x^*, x^{**}) = \sup_{(s, s^*) \in A} \{\langle s^*, x \rangle + \langle x^{**}, s^* \rangle - \langle s^*, s \rangle\} \text{ for all } (x^*, x^{**}) \in X^* \times X^{**}.$$

The function $\Phi_A : X \times X^* \to \overline{\mathbb{R}}$ has the following formula

$$\Phi_A(x, x^*) = \sup_{(s, s^*) \in A} \{\langle x^*, s \rangle + \langle s^*, x \rangle - \langle s^*, s \rangle\} \text{ for all } (x, x^*) \in X \times X^*,$$

that is Φ_A is the Fitzpatrick function of A (see Section 4.2). By applying the Fenchel-Moreau Theorem we obtain

$$\Phi_A^@ = \text{cl}_{s \times w^*}\,\text{co}(q + \delta_A) \tag{5. 11}$$

(the closure is taken with respect to the strong-weak* topology on $X \times X^*$). The function $^*\Theta_A : X \times X^* \to \overline{\mathbb{R}}$ has for all $(y, y^*) \in X \times X^*$ the following formula

$$^*\Theta_A(y, y^*) = \sup_{(x^*, x^{**}) \in X^* \times X^{**}} \{\langle x^*, y \rangle + \langle x^{**}, y^* \rangle - \Theta_A(x^*, x^{**})\}.$$

Example 5.4 As in [143, Remark 1] we consider E a nonreflexive Banach space, $X := E^*$ and $A := \{0\} \times \widehat{E}$, which is a monotone subset of $X \times X^*$. Let us notice that the relation $q + \delta_A = \delta_A$ is fulfilled. By applying Proposition 5.8(ii) we obtain $^*\Theta_A = \text{cl} \text{co} \, \delta_A = \delta_A$ (the closure is taken with respect to the strong topology of $X \times X^*$). Further, by using (5. 11) and the Goldstine Theorem we get $\Phi_A^{@} = \text{cl}_{s \times w^*} \text{co} \, \delta_A = \delta_{\{0\} \times E^{**}} \neq {}^*\Theta_A$ (the closure is considered with respect to the strong-weak* topology of $X \times X^*$).

Remark 5.9 Let us notice that the above example was given in [143] in order to show that in the nonreflexive case for a given monotone operator the family of representative functions which are strong-weak* lower semicontinuous does not coincide with the family of representative functions which are lower semicontinuous in the strong topology (see also the comments made in Remark 4.5). However, let us mention that the set A in Example 5.4 is not a maximal monotone subset of $X \times X^*$.

In the last part of the section we deal with another subfamily of $\mathbb{E}(A)$, namely the one of closed and *additive* enlargements. In this way we extend the results from [51, 134] to Banach SSD spaces.

Definition 5.8 *Let B be a Banach SSD space. We say that the multifunction $E : \mathbb{R}_+ \rightrightarrows B$ is additive if for all $\varepsilon_1, \varepsilon_2 \geq 0$ and $b^1 \in E(\varepsilon_1)$, $b^2 \in E(\varepsilon_2)$ one has*

$$q(b^1 - b^2) \geq -(\varepsilon_1 + \varepsilon_2).$$

In case $A \subseteq B$ is a maximally q-positive set we denote $\mathbb{E}_{ca}(A) := \{E \in \mathbb{E}_c(A) : E \text{ is additive}\}$.

We have the following characterization of the set $\mathbb{E}_{ca}(A)$.

Theorem 5.2 *Let B be a Banach SSD space, $A \subseteq B$ be a maximally q-positive set and $E \in \mathbb{E}_c(A)$. Then*

$$E \in \mathbb{E}_{ca}(A) \Leftrightarrow \Lambda_E^{@} \leq \Lambda_E.$$

Proof. Assume first that $E \in \mathbb{E}_{ca}(A)$ and take b^1, b^2 two arbitrary elements in B. By Proposition 5.5(iii) follows that $\lambda_E(b^1) \geq 0$ and $\lambda_E(b^2) \geq 0$. We claim that

$$q(b^1 - b^2) \geq -(\lambda_E(b^1) + \lambda_E(b^2)).$$

In case $\lambda_E(b^1) = +\infty$ or $\lambda_E(b^2) = +\infty$ (or both), this fact is obvious. If $\lambda_E(b^1)$ and $\lambda_E(b^2)$ are finite, the inequality above follows by using that (cf. Proposition 5.5(i)) $(b_1, \lambda_E(b^1)), (b_2, \lambda_E(b^2)) \in \text{epi}(\lambda_E) = G(E^{-1})$ and that E is additive. Consequently, for all $b^1, b^2 \in B$ we have (see also (5. 1))

$$\lambda_E(b^1) + q(b^1) \geq \lfloor b^1, b^2 \rfloor - (\lambda_E(b^2) + q(b^2)) \Leftrightarrow \Lambda_E(b^1) \geq \lfloor b^1, b^2 \rfloor - \Lambda_E(b^2).$$

This means that for all $b^1 \in B$, $\Lambda_E^{@}(b^1) \leq \Lambda_E(b^1)$.

Assume now that $\Lambda_E^{@} \leq \Lambda_E$ and take arbitrary $\varepsilon_1, \varepsilon_2 \geq 0$ and $b^1 \in E(\varepsilon_1)$, $b^2 \in E(\varepsilon_2)$. This means that $\lambda_E(b^1) \leq \varepsilon_1$ and $\lambda_E(b^2) \leq \varepsilon_2$. Since $\Lambda_E(b^1) \geq \Lambda_E^{@}(b^1)$, one has

$$\lambda_E(b^1) + q(b^1) \geq \lfloor b^1, b^2 \rfloor - (\lambda_E(b^2) + q(b^2)),$$

which yields (see also (5. 1))

$$q(b^1 - b^2) \geq -(\lambda_E(b^1) + \lambda_E(b^2)) \geq -(\varepsilon_1 + \varepsilon_2).$$

This concludes the proof.   □

The above characterization of the family of closed and additive enlargements associated to a maximally q-positive set yields in particular that this family is non-empty. This result is stated below.

Proposition 5.9 *Let B be a Banach SSD space and $A \subseteq B$ be a maximally q-positive set. Then $\overline{E_A} \in \mathbb{E}_{ca}(A)$, hence $\mathbb{E}_{ca}(A) \neq \emptyset$.*

Proof. By Proposition 5.8(iii) and Lemma 5.2(iii)-(iv) we have $(\Lambda_{\overline{E_A}})^@ = (^*\Theta_A)^@ = \Phi_A \leq {}^*\Theta_A = \Lambda_{\overline{E_A}}$. Theorem 5.2 guarantees that $\overline{E_A} \in \mathbb{E}_{ca}(A)$ and the proof is completed.   □

Theses

1. The convex optimization problem

$$(P_F) \quad \inf_{x \in X} \{f(x) + g(x)\}$$

is considered, where X is a separated locally convex space and $f, g : X \to \overline{\mathbb{R}}$ are proper and convex functions such that $\operatorname{dom} f \cap \operatorname{dom} g \neq \emptyset$. The Fenchel dual problem associated to (P_F) is

$$(D_F) \quad \sup_{x^* \in X^*} \{-f^*(-x^*) - g^*(x^*)\},$$

where X^* is the topological dual space of X and f^*, g^* are the Fenchel-Moreau conjugates of f, respectively g. We introduce some new regularity conditions expressed by means of the quasi interior and quasi-relative interior (a generalized interiority notion introduced by BORWEIN AND LEWIS in [14]). By using some separation theorems available for these notions, we prove that the regularity conditions introduced are sufficient for strong duality, the situation when the optimal objective values of the two problems coincide and the dual has an optimal solution. Moreover, these conditions offer an alternative for the case when the classical regularity conditions from the literature cannot be applied and this is illustrated by an example.

Further, corresponding regularity conditions are derived for the convex optimization problem

$$(P_F^A) \quad \inf_{x \in X} \{f(x) + (g \circ A)(x)\},$$

where X and Y are separated locally convex spaces with topological dual spaces X^* and Y^*, respectively, $A : X \to Y$ is a continuous linear mapping, $f : X \to \overline{\mathbb{R}}$ and $g : Y \to \overline{\mathbb{R}}$ are proper and convex functions such that $A(\operatorname{dom} f) \cap \operatorname{dom} g \neq \emptyset$. The Fenchel dual problem associated to (P_F^A) is

$$(D_F^A) \quad \sup_{y^* \in Y^*} \{-f^*(-A^*y^*) - g^*(y^*)\},$$

where $A^* : Y^* \to X^*$ is the adjoint operator of A (see [30]).

2. Consider the optimization problem with geometric and cone constraints

$$(P_L) \quad \inf_{\substack{x \in S \\ g(x) \in -C}} f(x),$$

where X is a topological vector space, Y is a separated locally convex space, S is a non-empty subset of X, $C \subseteq Y$ is a non-empty convex cone, $f : S \to \mathbb{R}$, $g : S \to Y$ and the constraint set $\mathcal{T} = \{x \in S : g(x) \in -C\}$ is assumed to be non-empty. The Lagrange dual problem associated to (P_L) is

$$(D_L) \quad \sup_{\lambda \in C^*} \inf_{x \in S} \{f(x) + \langle \lambda, g(x) \rangle\},$$

where $C^* = \{x^* \in X^* : \langle x^*, x \rangle \geq 0 \ \forall x \in C\}$ is the positive dual cone of C. The pair $(f,g) : S \to \mathbb{R} \times Y$, defined by $(f,g)(x) = (f(x), g(x))$ for all $x \in S$, is assumed to be convex-like with respect to the cone $\mathbb{R}_+ \times C \subseteq \mathbb{R} \times Y$, that is the set $(f,g)(S) + \mathbb{R}_+ \times C$ is convex.

We give regularity conditions for strong duality by means of the notions of quasi interior and quasi-relative interior. This is done by using an approach due to MAGNANTI (cf. [93]). He introduced a technique showing that Fenchel and Lagrange duality are "equivalent", in the sense that the classical Fenchel duality result can be derived from the classical Lagrange duality result and viceversa.

We also discuss some results recently given on this topic which are proved to have either superfluous or contradictory assumptions. Several examples are illustrating the theoretical considerations (see [26, 30]).

3. We give necessary and sufficient sequential optimality conditions for the general optimization problem

$$(P_\Phi) \quad \inf_{x \in X} \Phi(x, 0),$$

where X and Y are Banach spaces, X is reflexive and $\Phi : X \times Y \to \overline{\mathbb{R}}$ is a proper, convex and lower semicontinuous function fulfilling $0 \in \mathrm{pr}_Y(\mathrm{dom}\, \Phi)$. The sequential optimality conditions are expressed via the ε-subdifferential of the function Φ. By using a version of the Brøndsted-Rockafellar Theorem we derive sequential optimality conditions by means of the classical (convex) subdifferential.

We specialize these conditions to the optimization problem

$$(P_F^A) \quad \inf_{x \in X} \{f(x) + (g \circ A)(x)\},$$

obtaining in particular several sequential generalizations of the Pshenichnyi-Rockafellar Lemma and improving the sequential optimality conditions given by JEYAKUMAR AND WU in [86].

For an appropriate choice of the function Φ, we also get some sequential Lagrange multiplier conditions regarding the optimization problem with geometric and cone constraints

$$(P_L) \quad \inf_{\substack{x \in S \\ g(x) \in -C}} f(x),$$

showing that in the sequential optimality conditions given by THIBAULT in [138, Theorem 4.1] the hypothesis of normality for the cone C can be removed (see [28]).

By particularizing the general conditions given for Φ, we obtain sequential optimality conditions for the composed convex optimization problem

$$(P^{CC}) \quad \inf_{x \in X} \{f(x) + (g \circ h)(x)\},$$

where X is a reflexive Banach space, Y is a Banach space partially ordered by the non-empty convex cone $C \subseteq Y$, $f : X \to \overline{\mathbb{R}}$ is proper, convex and lower semicontinuous, $h : X \to Y^\bullet = Y \cup \{\infty_C\}$ is proper and C-convex and $g : Y^\bullet \to \overline{\mathbb{R}}$ is proper, convex and lower semicontinuous with $g(\infty_C) = +\infty$. We consider two cases, namely when h is C-epi-closed and g is C-increasing on $h(\mathrm{dom}\, h) + C$ (in which case Y is considered reflexive), respectively when $h : X \to Y$ is continuous and g is C-increasing on Y. We rediscover in this way (and improve in some conditions) several sequential optimality conditions given by THIBAULT in [138] (see [29]).

4. A necessary and sufficient closedness-type regularity condition is given in order to guarantee the following bivariate infimal convolution formula

$$(h_1 \square_2 h_2)^* = h_1^* \square_1 h_2^* \text{ and } h_1^* \square_1 h_2^* \text{ is exact,}$$

where X and Y are separated locally convex spaces and $h_1, h_2 : X \times Y \to \overline{\mathbb{R}}$ are proper, convex and lower semicontinuous functions such that $\mathrm{pr}_X(\mathrm{dom}\, h_1) \cap \mathrm{pr}_X(\mathrm{dom}\, h_2) \neq \emptyset$. Here, $h_1 \square_2 h_2 : X \times Y \to \overline{\mathbb{R}}$, $(h_1 \square_2 h_2)(x, y) = \inf\{h_1(x, u) + h_2(x, v) : u, v \in Y, u + v = y\}$ and $h_1^* \square_1 h_2^* : X^* \times Y^* \to \overline{\mathbb{R}}$, $(h_1^* \square_1 h_2^*)(x^*, y^*) = \inf\{h_1^*(u^*, y^*) + h_2^*(v^*, y^*) : u^*, v^* \in X^*, u^* + v^* = x^*\}$. As pointed out by many authors (see for example [113, 132]), such a formula is useful when dealing with the maximality of the sum of two maximal monotone operators in reflexive Banach spaces (see [22]).

5. Consider X a Banach space, $S, T : X \rightrightarrows X^*$ two maximal monotone operators with representative functions h_S, h_T, respectively, such that $\mathrm{pr}_X(\mathrm{dom}\, h_S) \cap \mathrm{pr}_X(\mathrm{dom}\, h_T) \neq \emptyset$ and the function $h : X \times X^* \to \overline{\mathbb{R}}$ defined by $h(x, x^*) = (h_S \square_2 h_T)^*(x^*, x)$ for all $(x, x^*) \in X \times X^*$. We give an application of the bivariate infimal convolution formula to enlargements of monotone operators and establish a necessary and sufficient condition for the following formula

$$(S + T)_h(\varepsilon, x) = \bigcup_{\substack{\varepsilon_1, \varepsilon_2 \geq 0 \\ \varepsilon_1 + \varepsilon_2 = \varepsilon}} \left(S_{h_S^*}(\varepsilon_1, x) + T_{h_T^*}(\varepsilon_2, x) \right) \ \forall \varepsilon \geq 0 \text{ and } \forall x \in X,$$

where for $M : X \rightrightarrows X^*$ a maximal monotone operator and $h_M : X \times X^* \to \overline{\mathbb{R}}$ a representative function of M, $M_{h_M} : \mathbb{R}_+ \times X \rightrightarrows X^*$, defined by $M_{h_M}(\varepsilon, x) = \{x^* \in X^* : h_M(x, x^*) \leq \varepsilon + \langle x^*, x \rangle\}$ is an enlargement of M. We generalize in this way the formula

$$\partial_\varepsilon(f + g)(x) = \bigcup_{\substack{\varepsilon_1, \varepsilon_2 \geq 0 \\ \varepsilon_1 + \varepsilon_2 = \varepsilon}} \left(\partial_{\varepsilon_1} f(x) + \partial_{\varepsilon_2} g(x) \right) \ \forall \varepsilon \geq 0 \text{ and } \forall x \in X,$$

which is equivalently characterized (in case f and g are proper, convex and lower semicontinuous functions such that $\mathrm{dom}\, f \cap \mathrm{dom}\, g \neq \emptyset$) by the condition $\mathrm{epi}\, f^* + \mathrm{epi}\, g^*$ is weak*-closed (see [22]).

6. We give an answer to the open problem posed in [44] concerning the characterization of the maximal monotone operators $S : X \rightrightarrows X^*$ which are fully enlargeable by S^{se}, the smallest element of $\mathbb{E}_c(S)$, a special family of enlargements associated to the maximal monotone operator S (see [24]).

7. Under a generalized interior-point condition we establish the weak*-closedness of the set $S_{h_S}(\varepsilon_1, x) + T_{h_T}(\varepsilon_2, x)$, where $S, T : X \rightrightarrows X^*$ are two maximal monotone operators with strong-weak* lower semicontinuous representative functions h_S and h_T, respectively. In case X is a reflexive Banach space, or X is Banach and S and T are of Gossez type (D), we improve a result given by García, Lassonde and Revalski in [67, Theorem 3.7 (1)] (see [24]).

8. Consider $(B, \lfloor \cdot, \cdot \rfloor)$ a symmetrically self-dual space (SSD space), a notion introduced and studied by Simons in [129]. The theory of monotone operators can be studied in this more general context. We introduce the notion of enlargement of a positive set in SSD spaces. To a maximally positive set A we associate a family of enlargements $\mathbb{E}(A)$ and characterize the smallest and biggest element in this family with respect to the inclusion relation. A one-to-one correspondence between the subfamily of closed enlargements of $\mathbb{E}(A)$

and the family of so-called representative functions of A is established in the framework of Banach SSD spaces. We show that the extremal elements of the latter family are two functions recently introduced and studied by SIMONS in [131]. In this way we extend to (Banach) SSD spaces some former results stated for monotone and maximally monotone sets in Banach spaces by BURACHIK AND SVAITER in [50, 51, 134] (see [23]).

Index of notation

$\forall$	for all
$\exists$	there exists (at least one)
$\mathbb{N}$	the set of positive integers $\{1, 2, ...\}$
$\mathbb{Z}$	the set of integer numbers
$\mathbb{R}$	the set of real numbers
$\overline{\mathbb{R}}$	the extended set of real numbers
$\mathbb{R}^m_+$	the non-negative orthant of $\mathbb{R}^m$
$\leq_C$	the partial ordering introduced by a non-empty convex cone C
C^*	the positive dual cone of the cone C
co(U)	the convex hull of the set U
cone(U)	the conic hull of the set U
coneco(U)	the convex conic hull of the set U
int(U)	the interior of the set U
ri(U)	the relative interior of the set U
core(U)	the algebraic interior of the set U
icr(U)	the relative algebraic interior of the set U
sqri(U)	the strong quasi-relative interior of the set U
qi(U)	the quasi interior of the set U
qri(U)	the quasi-relative interior of the set U
cl(U)	the closure of the set U
aff(U)	the affine hull of the set U
lin(U)	the linear subspace generated by the set U
dom f	the domain of the function f
epi f	the epigraph of the function f
epi$_C\, g$	the C-epigraph of g
co f	the convex hull of the function f
cl f	the lower semicontinuous hull of the function f
f^*	the Fenchel-Moreau conjugate of the function f
$\langle \cdot, \cdot \rangle$	the bilinear pairing between two vector spaces which are in duality

∂f the (convex) subdifferential of the function f

$\partial_\varepsilon f$ the ε-subdifferential of the function f

A^* the adjoint of the continuous linear mapping A

δ_U the indicator function of the set U

σ_U the support function of the set U

$v(P)$ the optimal objective value of the optimization problem (P)

$N_U(x)$ the normal cone to the set U at $x \in U$

$T_U(x)$ the contingent (Bouligand) cone to the set U at $x \in U$

X^* the topological dual space of the topological vector space X

$w(X, X^*)$ the weak topology on X induced by X^*

$w^*(X^*, X)$ the weak* topology on X^* induced by X

$\mathcal{R}$ the usual topology on $\mathbb{R}$

$S : X \rightrightarrows Y$ a set valued operator from X to Y

φ_S the Fitzpatrick function of the monotone operator $S : X \rightrightarrows X^*$

$(B, \lfloor \cdot, \cdot \rfloor)$ an SSD space

Bibliography

[1] M. Aït Mansour, A. Metrane, M. Théra, *Lower semicontinuous regularization for vector-valued mappings*, Journal of Global Optimization **35 (2)**, 283–309, 2006.

[2] H. Attouch, H. Brézis, *Duality for the sum of convex functions in general Banach spaces*, in: J.A. Barroso (ed.) Aspects of Mathematics and Its Applications, North-Holland Publishing Company, Amsterdam, 125–133, 1986.

[3] A. Auslender, M. Teboulle, *Asymptotic Cones and Functions in Optimization and Variational Inequalities*, Springer-Verlag, New York, 2003.

[4] M. Avriel, *Nonlinear Programming. Analysis and Methods*, Prentice-Hall Series in Automatic Computation, Prentice-Hall, Inc., Englewood Cliffs, N.J., 1976.

[5] H.H. Bauschke, *Fenchel duality, Fitzpatrick functions and the extension of firmly nonexpansive mappings*, Proceedings of the American Mathematical Society **135 (1)**, 135–139, 2007.

[6] H.H. Bauschke, D.A. McLaren, H.S. Sendov, *Fitzpatrick functions: inequalities, examples and remarks on a problem by S. Fitzpatrick*, Journal of Convex Analysis **13 (3-4)**, 499–523, 2006.

[7] H.H. Bauschke, X. Wang, L. Yao, *An answer to S. Simons' question on the maximal monotonicity of the sum of a maximal monotone linear operator and a normal cone operator*, Set-Valued and Variational Analysis **17 (2)**, 195–201, 2009.

[8] A. Ben-Israel, A. Ben-Tal, S. Zlobec, *Optimality conditions in convex programming*, in: A. Prékopa (ed.), Survey of Mathematical Programming, Proceedings of the 9th International Mathematical Programming Symposium (The Hungarian Academy of Sciences, Budapest and North-Holland, Amsterdam), **vol. 1**, 153–169, 1979.

[9] A. Ben-Tal, A. Ben-Israel, S. Zlobec, *Characterization of optimality in convex programming without a constraint qualification*, Journal of Optimization Theory and Applications **20 (4)**, 417–437, 1976.

[10] J.M. Borwein, *Maximal monotonicity via convex analysis*, Journal of Convex Analysis **13 (3-4)**, 561–586, 2006.

[11] J.M. Borwein, *Maximality of sums of two maximal monotone operators in general Banach space*, Proceedings of the American Mathematical Society **135 (12)**, 3917–3924, 2007.

[12] J.M. Borwein, R. Goebel, *Notions of relative interior in Banach spaces*, Journal of Mathematical Sciences **115 (4)**, 2542–2553, 2003.

[13] J.M. Borwein, V. Jeyakumar, A.S. Lewis, H. Wolkowicz, *Constrained approximation via convex programming*, Preprint, University of Waterloo, 1988.

[14] J.M. Borwein, A.S. Lewis, *Partially finite convex programming, part I: Quasi relative interiors and duality theory*, Mathematical Programming **57 (1)**, 15–48, 1992.

[15] J.M. Borwein, A.S. Lewis, *Convex Analysis and Nonlinear Optimization: Theory and Examples*, Second edition. CMS Books in Mathematics/Ouvrages de Mathématiques de la SMC, **3**. Springer-Verlag, New York, 2006.

[16] J.M. Borwein, Y. Lucet, B. Mordukhovich, *Compactly epi-Lipschitzian convex sets and functions in normed spaces*, Journal of Convex Analysis **7 (2)**, 375–393, 2000.

[17] J.M. Borwein, J.D. Vanderwerff, *Convex Functions: Constructions, Characterizations and Counterexamples*, Encyclopedia of Mathematics and its Applications **109**, Cambridge University Press, Cambridge, 2010.

[18] J.M. Borwein, H. Wolkowicz, *Characterizations of optimality for the abstract convex program with finite-dimensional range*, Journal of the Australian Mathematical Society Series A **30 (4)**, 390–411, 1981.

[19] J.M. Borwein, H. Wolkowicz, *Characterizations of optimality without constraint qualification for the abstract convex program*, Mathematical Programming Study **19**, 77–100, 1982.

[20] J.M. Borwein, Q.J. Zhu, *Techniques of Variational Analysis*, Springer-Verlag, New York, 2005.

[21] R.I. Boţ, *Conjugate Duality in Convex Optimization*, Lecture Notes in Economics and Mathematical Systems, Vol. **637**, Springer-Verlag Berlin Heidelberg, 2010.

[22] R.I. Boţ, E.R. Csetnek, *An application of the bivariate inf-convolution formula to enlargements of monotone operators*, Set-Valued Analysis **16 (7-8)**, 983–997, 2008.

[23] R.I. Boţ, E.R. Csetnek, *Enlargements of positive sets*, Journal of Mathematical Analysis and Applications **356 (1)**, 328–337, 2009.

[24] R.I. Boţ, E.R. Csetnek, *On two properties of enlargements of maximal monotone operators*, Journal of Convex Analysis **16 (3-4)**, 713–725, 2009.

[25] R.I. Boţ, E.R. Csetnek, *Regularity conditions via generalized interiority notions in convex optimization: new achievements and their relation to some classical statements*, Optimization, to appear, arXiv:0906.0453v1, posted 2 June, 2009.

[26] R.I. Boţ, E.R. Csetnek, A. Moldovan, *Revisiting some duality theorems via the quasirelative interior in convex optimization*, Journal of Optimization Theory and Applications **139 (1)**, 67–84, 2008.

[27] R.I. Boţ, E.R. Csetnek, G. Wanka, *A new condition for maximal monotonicity via representative functions*, Nonlinear Analysis: Theory, Methods & Applications **67 (8)**, 2390–2402, 2007.

[28] R.I. Boţ, E.R. Csetnek, G. Wanka, *Sequential optimality conditions in convex programming via perturbation approach*, Journal of Convex Analysis **15 (1)**, 149–164, 2008.

[29] R.I. Boţ, E.R. Csetnek, G. Wanka, *Sequential optimality conditions for composed convex optimization problems*, Journal of Mathematical Analysis and Applications **342 (2)**, 1015–1025, 2008.

[30] R.I. Boţ, E.R. Csetnek, G. Wanka, *Regularity conditions via quasi-relative interior in convex programming*, SIAM Journal on Optimization **19 (1)**, 217–233, 2008.

[31] R.I. Boţ, A. Grad, G. Wanka, *Sequential characterization of solutions in convex composite programming and applications to vector optimization*, Journal of Industrial and Management Optimization **4 (4)**, 767–782, 2008.

[32] R.I. Boţ, S.-M. Grad, G. Wanka, *Maximal monotonicity for the precomposition with a linear operator*, SIAM Journal on Optimization **17 (4)**, 1239–1252, 2006.

[33] R.I. Boţ, S.-M. Grad, G. Wanka, *Weaker constraint qualifications in maximal monotonicity*, Numerical Functional Analysis and Optimization **28 (1-2)**, 27–41, 2007.

[34] R.I. Boţ, S.-M. Grad, G. Wanka, *On strong and total Lagrange duality for convex optimization problems*, Journal of Mathematical Analysis and Applications **337 (2)**, 1315–1325, 2008.

[35] R.I. Boţ, S.-M. Grad, G. Wanka, *Generalized Moreau-Rockafellar results for composed convex functions*, Optimization **58 (7)**, 917–933, 2009.

[36] R.I. Boţ, S.-M. Grad, G. Wanka, *Duality in Vector Optimization*, Springer-Verlag, Berlin, 2009.

[37] R.I. Boţ, I.B. Hodrea, G. Wanka, *ε-optimality conditions for composed convex optimization problems*, Journal of Approximation Theory **153 (1)**, 108–121, 2008.

[38] R.I. Boţ, G. Wanka, *An alternative formulation for a new closed cone constraint qualification*, Nonlinear Analysis: Theory, Methods & Applications **64 (6)**, 1367–1381, 2006.

[39] R.I. Boţ, G. Wanka, *A weaker regularity condition for subdifferential calculus and Fenchel duality in infinite dimensional spaces*, Nonlinear Analysis: Theory, Methods & Applications **64 (12)**, 2787–2804, 2006.

[40] A. Brøndsted, *Conjugate convex functions in topological vector spaces*, Matematiskfysiske Meddelelser udgivet af det Kongelige Danske Videnskabernes Selskab **34 (2)**, 1–27, 1964.

[41] A. Brøndsted, R.T. Rockafellar, *On the subdifferentiability of convex functions*, Proceedings of the American Mathematical Society **16**, 605–611, 1965.

[42] F.E. Browder, *Nonlinear maximal monotone operators in Banach spaces*, Mathematische Annalen **175**, 89–113, 1968.

[43] R.S. Burachik, S. Fitzpatrick, *On a family of convex functions associated to subdifferentials*, Journal of Nonlinear and Convex Analysis **6 (1)**, 165–171, 2005.

[44] R.S. Burachik, A.N. Iusem, *On non-enlargeable and fully enlargeable monotone operators*, Journal of Convex Analysis **13 (3-4)**, 603–622, 2006.

[45] R.S. Burachik, A.N. Iusem, *Set-valued Mappings and Enlargements of Monotone Operators*, Springer Optimization and Its Applications **8**, Springer, New York, 2008.

[46] R.S. Burachik, A.N. Iusem, B.F. Svaiter, *Enlargement of monotone operators with applications to variational inequalities*, Set-Valued Analysis **5 (2)**, 159–180, 1997.

[47] R.S. Burachik, V. Jeyakumar, *A new geometric condition for Fenchel's duality in infinite dimensional spaces*, Mathematical Programming **104 (2-3)**, 229–233, 2005.

[48] R.S. Burachik, V. Jeyakumar, Z.-Y. Wu, *Necessary and sufficient conditions for stable conjugate duality*, Nonlinear Analysis: Theory, Methods & Applications **64 (9)**, 1998–2006, 2006.

[49] R.S. Burachik, C.A. Sagastizábal, B.F. Svaiter, *ε-enlargements of maximal monotone operators: theory and applications*, In: M. Fukushima and L. Qi (eds), Reformulation: Nonsmooth, Piecewise Smooth, Semismooth and Smoothing Methods (Lausanne, 1997), Appl. Optim. **22**, Kluwer Acad. Publ., Dordrecht, 25–43, 1999.

[50] R.S. Burachik, B.F. Svaiter, *ε-enlargements of maximal monotone operators in Banach spaces*, Set-Valued Analysis **7 (2)**, 117–132, 1999.

[51] R.S. Burachik, B.F. Svaiter, *Maximal monotone operators, convex functions and a special family of enlargements*, Set-Valued Analysis **10 (4)**, 297–316, 2002.

[52] R.S. Burachik, B.F. Svaiter, *Maximal monotonicity, conjugation and duality product*, Proceedings of the American Mathematical Society **131 (8)**, 2379–2383, 2003.

[53] R.S. Burachik, B.F. Svaiter, *Operating enlargements of monotone operators: new connections with convex functions*, Pacific Journal of Optimization **2 (3)**, 425–445, 2006.

[54] J.V. Burke, S. Deng, *Weak sharp minima revisited, part II: application to linear regularity and error bounds*, Mathematical Programming Series B **104 (2-3)**, 235–261, 2005.

[55] F. Cammaroto, B. Di Bella, *Separation theorem based on the quasirelative interior and application to duality theory*, Journal of Optimization Theory and Applications **125 (1)**, 223–229, 2005.

[56] F. Cammaroto, B. Di Bella, *On a separation theorem involving the quasirelative interior*, Proceedings of the Edinburg Mathematical Society (2) **50 (3)**, 605–610, 2007.

[57] C. Combari, M. Laghdir, L. Thibault, *Sous-différentiels de fonctions convexes composées*, Annales des Sciences Mathématiques du Québec **18 (2)**, 119–148, 1994.

[58] B.D. Craven, S. Zlobec, *Complete characterization of optimality of convex programming in Banach spaces*, Applicable Analysis **11 (1)**, 61–78, 1980.

[59] P. Daniele, *Lagrange multipliers and infinite-dimensional equilibrium problems,* Journal of Global Optimization **40 (1-3)**, 65–70, 2008.

[60] P. Daniele, S. Giuffrè, *General infinite dimensional duality and applications to evolutionary network equilibrium problems*, Optimization Letters **1** (**3**), 227–243, 2007.

[61] P. Daniele, S. Giuffrè, G. Idone, A. Maugeri, *Infinite dimensional duality and applications*, Mathematische Annalen **339** (**1**), 221–239, 2007.

[62] I. Ekeland, R. Temam, *Convex Analysis and Variational Problems*, North-Holland Publishing Company, Amsterdam, 1976.

[63] M. Fabian, P. Habala, P. Hájek, V. Montesinos Santaluca, J. Pelant, V. Zizler, *Functional Analysis and Infinite-Dimensional Geometry*, CMS Books in Mathematics/Ouvrages de Mathématiques de la SMC **8**, Springer-Verlag, New York, 2001.

[64] W. Fenchel, *On conjugate convex functions*, Canadian Journal of Mathematics **1**, 73–77, 1949.

[65] S. Fitzpatrick, *Representing monotone operators by convex functions*, in: Workshop/Miniconference on Functional Analysis and Optimization (Canberra, 1988), Proceedings of the Centre for Mathematical Analysis **20**, Australian National University, Canberra, 59–65, 1988.

[66] D. Gale, H.W. Kuhn, A.W. Tucker, *Linear Programming and the Theory of Games*, in: T.C. Koopman (ed.), Activity Analysis of Production and Allocation, John Wiley & Sons, Inc., New York, 317–329, 1951.

[67] Y. García, M. Lassonde, J.P. Revalski, *Extended sums and extended compositions of monotone operators*, Journal of Convex Analysis **13** (**3-4**), 721–738, 2006.

[68] F. Giannessi, *Constrained Optimization and Image Space Analysis, Vol. 1. Separation of Sets and Optimality Conditions*, Mathematical Concepts and Methods in Science and Engineering **49**, Springer, New York, 2005.

[69] S. Giuffrè, G. Idone, A. Maugeri, *Duality theory and optimality conditions for generalized complementarity problems*, Nonlinear Analysis: Theory, Methods & Applications **63** (**5-7**), e1655–e1664, 2005.

[70] S. Giuffrè, G. Idone, A. Maugeri, *Optimality conditions for generalized complementarity problems*, in: F. Giannessi, A. Maugeri (eds.), Variational Analysis and Applications, Nonconvex Optimization and its Applications **79**, Springer, New York, 465–475, 2005.

[71] M.S. Gowda, M. Teboulle, *A comparison of constraint qualifications in infinite-dimensional convex programming*, SIAM Journal on Control and Optimization **28** (**4**), 925–935, 1990.

[72] A. Grad, *Quasi-relative interior-type constraint qualifications ensuring strong Lagrange duality for optimization problems with cone and affine constraints*, Journal of Mathematical Analysis and Applications **361** (**1**), 86–95, 2010.

[73] A. Hantoute, M.A. López, C. Zălinescu, *Subdifferential calculus rules in convex analysis: a unifying approach via pointwise supremum functions*, SIAM Journal on Optimization **19** (**2**), 863–882, 2008.

[74] J.-B. Hiriart-Urruty, *ε-subdifferential calculus*, in: J.-P. Aubin, R.B. Vinter (eds.), Convex Analysis and Optimization, Research Notes in Mathematics **57**, Pitman, Boston, 43–92, 1982.

[75] J.-B. Hiriart-Urruty, C. Lemaréchal, *Convex Analysis and Minimization Algorithms I: Fundamentals*, Springer-Verlag, Berlin, 1993.

[76] J.-B. Hiriart-Urruty, C. Lemaréchal, *Convex Analysis and Minimization Algorithms II: Advanced theory and bundle methods*, Springer-Verlag, Berlin, 1993.

[77] J.-B. Hiriart-Urruty, C. Lemaréchal, *Fundamentals of Convex Analysis*, Springer-Verlag, Berlin, 2001.

[78] R.B. Holmes, *Geometric Functional Analysis and its Applications*, Springer-Verlag, Berlin, 1975.

[79] A. Ioffe, *Three theorems on subdifferentiation of convex integral functionals*, Journal of Convex Analysis **13 (3-4)**, 759–772, 2006.

[80] J. Jahn, *Introduction to the Theory of Nonlinear Optimization*, Springer-Verlag, Berlin, 1996.

[81] V. Jeyakumar, *Asymptotic dual conditions characterizing optimality for infinite convex programs*, Journal of Optimization Theory and Applications **93 (1)**, 153–165, 1997.

[82] V. Jeyakumar, N. Dinh, G. M. Lee, *A new closed cone constraint qualification for convex optimization*, Applied Mathematics Report AMR 04/8, University of New South Wales, Sydney, Australia, 2004.

[83] V. Jeyakumar, G.M. Lee, N. Dinh, *New sequential Lagrange multiplier conditions characterizing optimality without constraint qualifications for convex programs*, SIAM Journal on Optimization **14 (2)**, 534–547, 2003.

[84] V. Jeyakumar, W. Song, N. Dinh, G.M. Lee, *Stable strong duality in convex optimization*, Applied Mathematics Report AMR 05/22, University of New South Wales, Sydney, Australia, 2005.

[85] V. Jeyakumar, H. Wolkowicz, *Generalizations of Slater's constraint qualification for infinite convex programs*, Mathematical Programming Series B **57 (1)**, 85–101, 1992.

[86] V. Jeyakumar, Z.Y. Wu, *A qualification free sequential Pshenichnyi-Rockafellar Lemma and convex semidefinite programming*, Journal of Convex Analysis **13 (3-4)**, 773–784, 2006.

[87] V. Jeyakumar, Z.Y. Wu, *A dual criterion for maximal monotonicity of composition operators*, Set-Valued Analysis **15 (3)**, 265–273, 2007

[88] V. Jeyakumar, Z.Y. Wu, G.M. Lee, N. Dinh, *Liberating the subgradient optimality conditions from constraint qualifications*, Journal of Global Optimization **36 (1)**, 127–137, 2006.

[89] V. Jeyakumar, A. Zaffaroni, *Asymptotic conditions for weak and proper optimality in infinite dimensional convex vector optimization*, Numerical Functional Analysis and Optimization **17 (3-4)**, 323–343, 1996.

[90] C. Li, D. Fang, G. López, M.A. López, *Stable and total Fenchel duality for convex optimization problems in locally convex spaces*, SIAM Journal on Optimization **20 (2)**, 1032–1051, 2009.

[91] M.A. Limber, R.K. Goodrich, *Quasi interiors, Lagrange multipliers, and L^p spectral estimation with lattice bounds*, Journal of Optimization Theory and Applications **78 (1)**, 143–161, 1993.

[92] D.T. Luc, *Theory of Vector Optimization*, Springer-Verlag, Berlin, 1989.

[93] T.L. Magnanti, *Fenchel and Lagrange duality are equivalent*, Mathematical Programming **7**, 253–258, 1974.

[94] M. Marques Alves, B.F. Svaiter, *Brønsted-Rockafellar property and maximality of monotone operators representable by convex functions in non-reflexive Banach spaces*, Journal of Convex Analysis **15 (4)**, 693–706, 2008.

[95] M. Marques Alves, B.F. Svaiter, *A new old class of maximal monotone operators*, Journal of Convex Analysis **16 (3-4)**, 881–890, 2009.

[96] M. Marques Alves, B.F. Svaiter, *On Gossez type (D) maximal monotone operators*, Journal of Convex Analysis **17 (3-4)**, 2010.

[97] J.E. Martínez-Legaz, *On maximally q-positive sets*, Journal of Convex Analysis **16 (3-4)**, 891–898, 2009.

[98] J.E. Martínez-Legaz, B.F. Svaiter, *Monotone operators representable by l.s.c. convex functions*, Set-Valued Analysis **13 (1)**, 21–46, 2005.

[99] J.E. Martínez-Legaz, M. Théra, *ε-subdifferentials in terms of subdifferentials*, Set-Valued Analysis **4 (4)**, 327–332, 1996.

[100] J.E. Martínez-Legaz, M. Théra, *A convex representation of maximal monotone operators*, Journal of Nonlinear and Convex Analysis **2 (2)**, 243–247, 2001.

[101] A. Maugeri, F. Raciti, *On general infinite dimensional complementarity problems*, Optimization Letters **2 (1)**, 71–90, 2008.

[102] G.J. Minty, *Monotone (nonlinear) operators in Hilbert space*, Duke Mathematical Journal **29**, 341–346, 1962.

[103] A. Moldovan, *On Regularity for Constrained Extremum Problems*, PhD Thesis, Faculty of Mathematics, University of Pisa, 2008.

[104] J.J. Moreau, *Fonctions convexes en dualité*, (multigraph), Faculté des Sciences, Séminaires de Mathématiques, Université de Montpellier, Montpellier, 1962.

[105] J.J. Moreau, *Fonctionnelles convexes*, Seminaire sur les Équation aux Dérivées Partielles, Collége de France, Paris, 1967.

[106] J.-P. Penot, *Subdifferential calculus without qualification assumptions*, Journal of Convex Analysis **3 (2)**, 207–219, 1996.

[107] J.-P. Penot, *Is convexity useful for the study of monotonicity?*, in: R.P. Agarwal, D. O'Regan (eds.), "Nonlinear Analysis and Applications", Kluwer, Dordrecht, **vol. 1, 2**, 807–822, 2003.

[108] J.-P. Penot, *A representation of maximal monotone operators by closed convex functions and its impact on calculus rules*, Comptes Rendus Mathématique. Académie des Sciences. Paris **338 (11)**, 853–858, 2004.

[109] J.-P. Penot, *The relevance of convex analysis for the study of monotonicity*, Nonlinear Analysis: Theory, Methods & Applications **58** **(7-8)**, 855–871, 2004.

[110] J.-P. Penot, *Positive sets, conservative sets and dissipative sets*, Journal of Convex Analysis **16** **(3-4)**, 973–986, 2009.

[111] J.-P. Penot, M. Théra, *Semi-continuous mappings in general topology*, Archiv der Mathematik **38** **(2)**, 158-166, 1982.

[112] J.-P. Penot, C. Zălinescu, *Bounded convergence for perturbed minimization problems*, Optimization **53** **(5-6)**, 625–640, 2004.

[113] J.-P. Penot, C. Zălinescu, *Some problems about the representation of monotone operators by convex functions*, The ANZIAM Journal (The Australian & New Zealand Industrial and Applied Mathematics Journal) **47** **(1)**, 1–20, 2005.

[114] R.R. Phelps, *Lectures on maximal monotone operators*, Extracta Mathematicae **12** **(3)**, 193–230, 1997.

[115] J-Ch. Pomerol, *Contribution à la programmation mathématique: Existence des multiplicateurs de Lagrange et stabilité*, PhD Thesis, P. and M. Curie University, Paris, 1980.

[116] J. Ponstein, *Approaches to the Theory of Optimization*, Cambridge Tracts in Mathematics **77**, Cambridge University Press, Cambridge-New York, 1980.

[117] T. Precupanu, *Closedness conditions for the optimality of a family of non-convex optimization problems*, Mathematische Operationsforschung und Statistik Series Optimization **15** **(3)**, 339–346, 1984.

[118] B.N. Pshenichnyi, *Necessary Conditions for an Extremum*, Marcel Dekker, Inc., New York, 1971.

[119] J.P. Revalski, M. Théra, *Enlargements and sums of monotone operators*, Nonlinear Analysis: Theory, Methods & Applications **48** **(4)**, 505–519, 2002.

[120] R.T. Rockafellar, *Duality theorems for convex functions*, Bulletin of the American Mathematical Society **70**, 189–192, 1964.

[121] R.T. Rockafellar, *Extension of Fenchel's duality theorem for convex functions*, Duke Mathematical Journal **33** **(1)**, 81–89, 1966.

[122] R.T. Rockafellar, *Convex Analysis*, Princeton University Press, Princeton, 1970.

[123] R.T. Rockafellar, *Conjugate duality and optimization*, Conference Board of the Mathematical Sciences Regional Conference Series in Applied Mathematics **16**, Society for Industrial and Aplied Mathematics, Philadelphia, 1974.

[124] R.T. Rockafellar, *On the maximal monotonicity of subdifferential mappings*, Pacific Journal of Mathematics **33** **(1)**, 209–216, 1970.

[125] R.T. Rockafellar, *On the maximality of sums of nonlinear monotone operators*, Transactions of the American Mathematical Society **149**, 75–88, 1970.

[126] B. Rodrigues, *The Fenchel duality theorem in Fréchet spaces*, Optimization **21** **(1)**, 13–22, 1990.

[127] W. Rudin, *Functional Analysis*, Second edition, International Series in Pure and Applied Mathematics, McGraw-Hill, Inc., New York, 1991.

[128] S. Simons, *Minimax and Monotonicity*, Springer-Verlag, Berlin, 1998.

[129] S. Simons, *Positive sets and monotone sets*, Journal of Convex Analysis **14** (2), 297–317, 2007.

[130] S. Simons, *From Hahn-Banach to Monotonicity*, Springer-Verlag, Berlin, 2008.

[131] S. Simons, *Nonreflexive Banach SSD spaces*, Preprint, arXiv:0810.4579v2, posted 3 November, 2008.

[132] S. Simons, C. Zălinescu, *Fenchel duality, Fitzpatrick functions and maximal monotonicity*, Journal of Nonlinear and Convex Analysis **6** (1), 1–22, 2005.

[133] T. Strömberg, *The operation of infimal convolution*, Dissertationes Mathematicae **352**, 1996.

[134] B.F. Svaiter, *A family of enlargements of maximal monotone operators*, Set-Valued Analysis **8** (4), 311–328, 2000.

[135] T. Tanaka, *Generalized semicontinuity and existence theorems for cone saddle points*, Applied Mathematics and Optimization **36** (3), 313-322, 1997.

[136] T. Tanaka, D. Kuroiwa, *The convexity of A and B assures* int $A+B =$ int$(A+B)$, Applied Mathematics Letters **6** (1), 83–86, 1993.

[137] L. Thibault, *A short note on sequential convex subdifferential calculus* (unpublished paper), 1994.

[138] L. Thibault, *Sequential convex subdifferential calculus and sequential Lagrange multipliers*, SIAM Journal on Control and Optimization **35** (4), 1434–1444, 1997.

[139] L. Thibault, *Limiting convex subdifferential calculus with applications to integration and maximal monotonicity of subdifferential*, in: M. Théra (ed.), Constructive, Experimental, and Nonlinear Analysis, CMS Conference Proceedings **27**, American Mathematical Society, Providence, 279–289, 2000.

[140] D. Torralba, *Convergence épigraphique et changements d'échelle en analyse variationnelle et optimisation. Applications aux transitions de phases et à la méthode barrière logarithmique*, PhD Thesis, Université Montpellier II, 1996.

[141] A. Verona, M.E. Verona, *Regular maximal monotone operators and the sum theorem*, Journal of Convex Analysis **7** (1), 115–128, 2000.

[142] M.D. Voisei, *Calculus rules for maximal monotone operators in general Banach spaces*, Journal of Convex Analysis **15** (1), 73–85, 2008.

[143] M.D. Voisei, C. Zălinescu: *Linear monotone subspaces of locally convex spaces*, Set-Valued and Variational Analysis **18** (1), 29–55, 2010.

[144] M.D. Voisei, C. Zălinescu, *Strongly-representable monotone operators*, Journal of Convex Analysis **16** (3-4), 1011–1033, 2009.

[145] M. Volle, *On the subdifferential of an upper envelope of convex functions*, Acta Mathematica Vietnamica **19** (2), 137–148, 1994.

[146] D. Zagrodny, *The convexity of the closure of the domain and the range of a maximal monotone multifunction of type NI*, Set-Valued Analysis **16 (5-6)**, 759–783, 2008.

[147] C. Zălinescu, *Solvability results for sublinear functions and operators*, Zeitschrift für Operations Research Series A-B **31 (3)**, A79–A101, 1987.

[148] C. Zălinescu, *A comparison of constraint qualifications in infinite-dimensional convex programming revisited*, Journal of Australian Mathematical Society Series B **40 (3)**, 353–378, 1999.

[149] C. Zălinescu, *Convex Analysis in General Vector Spaces*, World Scientific, Singapore, 2002.

[150] C. Zălinescu, *Slice convergence for some classes of convex functions*, Journal of Nonlinear and Convex Analysis **4 (2)**, 185–214, 2003.

[151] C. Zălinescu, *A new proof of the maximal monotonicity of the sum using the Fitzpatrick function*, in: F. Giannessi, A. Maugeri (eds.), Variational Analysis and Applications, Nonconvex Optimization and its Applications **79**, Springer, New York, 1159–1172, 2005.